U0322176

NUAA Excellent Textbook for Postgraduates

南京航空航天大学研究生系列精品教材

Optical Measurement Mechanics

光 测 力 学

（双语版）

Kaifu WANG

王开福 著

科 学 出 版 社

北 京

内 容 简 介

The main contents of this book include holography and holographic inter-ferometry, speckle photography and speckle interferometry, geometric moiré and moiré interferometry, phase-shifting interferometry and phase unwrapping, discrete transformation and low-pass filtering, digital holography and digital holographic interferometry, digital speckle interferometry and digital speckle shearing interferometry, digital image correlation and particle image velocimetry, etc.

This book can be used as the English, Chinese, or bilingual textbook of optical measurement mechanics for postgraduate students majoring in mechanics, aeronautical engineering, mechanical engineering, optical engineering, civil engineering, etc.

本书主要内容包括全息照相与全息干涉、散斑照相与散斑干涉、几何云纹与云纹干涉、相移干涉与相位展开、离散变换与低通滤波、数字全息照相与数字全息干涉、数字散斑干涉与数字散斑剪切干涉、数字图像相关与粒子图像测速等。

本书可用作力学、航空工程、机械工程、光学工程、土木工程等学科专业研究生的光测力学英文、中文或双语教材。

图书在版编目(CIP)数据

光测力学＝Optical Measurement Mechanics：双语版：汉、英 / 王开福著.
—北京：科学出版社，2016.12
ISBN 978-7-03-051004-4

Ⅰ. ①光… Ⅱ. ①王… Ⅲ. ①光测力学－双语教学－高等学校－教材－汉、英 Ⅳ. ①O348.1

中国版本图书馆 CIP 数据核字(2016)第 298064 号

责任编辑：余 江 张丽花 / 责任校对：郭瑞芝
责任印制：徐晓晨 / 封面设计：迷底书装

科 学 出 版 社 出版
北京东黄城根北街 16 号
邮政编码：100717
http://www.sciencep.com

北京厚诚则铭印刷科技有限公司 印刷
科学出版社发行 各地新华书店经销

*

2016 年 12 月第 一 版 开本：787×1092 1/16
2018 年 1 月第二次印刷 印张：12 1/2
字数：290 000

定价：58.00 元

(如有印装质量问题，我社负责调换)

Preface

Optical measurement mechanics is an important degree course for postgraduate students majoring in mechanics, aeronautical engineering, mechanical engineering, optical engineering, civil engineering, etc. This book is intended to provide these postgraduate students with the fundamental theory and engineering application of optical measurement mechanics.

This book is written by using English and Chinese, respectively; the English part can be used as an English textbook of optical measurement mechanics, while the Chinese part can be used as a Chinese textbook of optical measurement mechanics. A combination of these two parts can be used as a bilingual textbook of optical measurement mechanics.

The main contents of this book include holography and holographic interferometry, speckle photography and speckle interferometry, geometric moiré and moiré interferometry, phase-shifting interferometry and phase unwrapping, discrete transformation and low-pass filtering, digital holography and digital holographic interferometry, digital speckle interferometry and digital speckle shearing interferometry, digital image correlation and particle image velocimetry, etc.

Kaifu WANG

Nanjing.June 2016

前　　言

光测力学是力学、航空工程、机械工程、光学工程和土木工程等专业研究生的重要学位课。本书旨在向这些研究生传授光测力学的基础理论与工程应用。

本书由英语和汉语分别编写：英语部分可作为光测力学的英文教材，而汉语部分则可作为光测力学的中文教材。两者结合可作为光测力学的双语教材。

本书主要内容包括全息照相与全息干涉、散斑照相与散斑干涉、几何云纹与云纹干涉、相移干涉与相位展开、离散变换与低通滤波、数字全息照相与数字全息干涉、数字散斑干涉与数字散斑剪切干涉、数字图像相关与粒子图像测速等。

王开福

2016 年 6 月于南京

Contents

目　　录

Chapter 1 Optical Measurement Mechanics Fundamentals

Optical measurement mechanics (or photomechanics) is an experimental interdiscipline related to optics and mechanics. Optical measurement mechanics can be used for solving mechanical problems, such as deformation measurement, vibration analysis, non-destructive test, etc., by employing optical techniques, such as holographic interferometry, speckle interferometry, moiré interferometry, etc.

1.1 Optics

Optics is the branch of physics that is concerned with light and vision and deals chiefly with the generation, propagation, and detection of electromagnetic radiation having wavelengths greater than X-ray and shorter than microwaves.

Optics usually describes the behavior and property of light. Light is electromagnetic radiation within a certain portion of the electromagnetic spectrum and usually refers to visible light which is visible to the human eye and is responsible for the sense of sight. Visible light has wavelengths in the range from about 400 (violet) to 760 (red) nanometre between infrared (with longer wavelengths) and ultraviolet (with shorter wavelengths).

Often, infrared and ultraviolet are also called light. Infrared light has the range of invisible radiation wavelengths from about 760 nanometers, just greater than red in the visible spectrum, to 1 millimeter, on the border of the microwave region. Ultraviolet light has the range of invisible radiation wavelengths from about 4 nanometers, on the border of the X-ray region, to about 400 nanometers, just beyond the violet in the visible spectrum.

1.1.1 Geometrical Optics

Most optical phenomena can be accounted for using the classical electromagnetic description of light. However, a complete electromagnetic description of light is often difficult to apply in practice. Therefore, simplified models are usually utilized in optics. One of the commonly used models is called geometrical optics, or ray optics. Geometrical optics treats light as a collection of light rays and describes light propagation in terms of light rays. A light ray is a straight or curved line that is perpendicular to the wavefront of light and is therefore collinear with the wave vector. A slightly more rigorous definition of light ray follows from Fermat's principle, which states that the path taken between two points by a ray of light is the path that can be traversed in the least time. The light ray in geometrical optics is useful in approximating the paths along which light propagates in certain classes of media.

The simplifying assumptions in geometrical optics mainly include that light rays: ①propagate in rectilinear paths as they travel in a homogeneous medium; ②bend at the interface between two dissimilar media; ③follow curved paths in a medium in which the refractive index changes. These simplifications are excellent approximations when the wavelength is small compared to the size of media with which the light interacts.

Geometrical optics is particularly useful in describing geometrical aspects of imaging, including optical aberrations. It is often simplified by making the paraxial approximation, or small angle approximation. The mathematical behavior then becomes linear, allowing optical systems to be described by simple matrices. This leads to the technique of paraxial ray tracing, which is used to find basic properties of optical systems, such as image positions.

1. Reflection and Refraction

Glossy surfaces such as mirrors reflect light in a simple, predictable way. This allows for production of reflected images that can be associated with an actual (real) or extrapolated (virtual) location in space. With such surfaces, the direction of the reflected ray is determined by the angle the incident ray makes with the surface normal, a line perpendicular to the surface at the point where the ray hits. The incident and reflected rays lie in a single plane, and the angle between the reflected ray and the surface normal is the same as that between the incident ray and the normal, i.e.,

$$\alpha = \beta \tag{1.1}$$

where α and β are respectively the incidence and reflection angles, as shown in Fig. 1.1. This is known as the law of reflection.

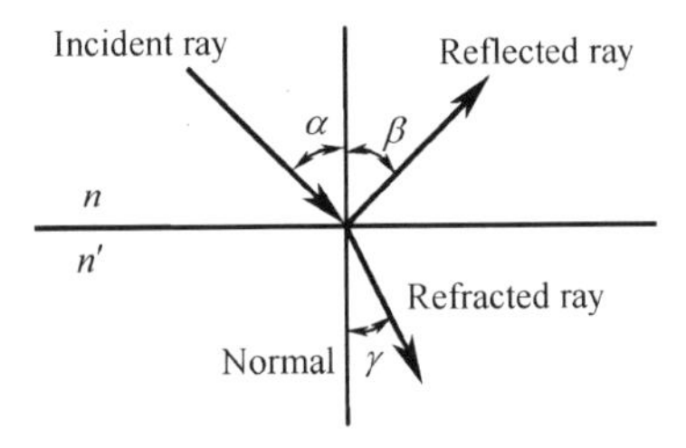

Fig. 1.1 Reflection and refraction

For flat mirrors, the law of reflection implies that images of objects are upright and the same distance behind the mirror as the objects are in front of the mirror. The image size is the same as the object size, i.e., the magnification of a flat mirror is equal to one. The law also implies that mirror images are parity inverted, which is perceived as a left-right inversion.

Mirrors with curved surfaces can be modeled by ray tracing and using the law of reflection at each point on the surface. For mirrors with parabolic surfaces, parallel rays incident on the mirror produce reflected rays that converge at a common focus. Other curved surfaces may also focus light, but with aberrations due to the diverging shape causing the focus to be smeared out in space. In particular, spherical mirrors exhibit spherical aberration. Curved mirrors can form images with magnification greater than or less than one, and the image can be upright or inverted. An upright image formed by reflection in a mirror is always virtual, while an inverted image is real and can be projected onto a screen.

Refraction occurs when light travels through an area of space that has a changing index of refraction. The simplest case of refraction occurs when there is an interface between a uniform medium with index of refraction n and another uniform medium with index of refraction n', as shown in Fig. 1.1. In such a situation, the law of refraction describes the resulting deflection of the light ray:

$$\frac{\sin\alpha}{\sin\gamma}=\frac{n'}{n} \tag{1.2}$$

where α and γ are the angles between the normal of the interface and the incident and refracted waves, respectively, i.e., the incidence and refraction angles.

The law of refraction can be used to predict the deflection of light rays as they pass through linear media as long as the indexes of refraction and the geometry of the media are known. For example, the propagation of light through a prism results in the light ray being deflected depending on the shape and orientation of the prism. Additionally, since different frequencies of light have slightly different indexes of refraction in most media, refraction can be used to produce dispersion spectra that appear as rainbows.

Some media have an index of refraction which varies gradually with position and, thus, light rays curve through the medium rather than travel in straight lines. This effect is the reason for the formation of mirages which can be observed on hot days where the changing index of refraction of the air causes the light rays to bend.

Various consequences of the law of refraction include the fact that for light rays traveling from a medium with a high index of refraction (i.e., optically denser medium) to a medium with a low index of refraction (i.e., optically thinner medium), it is possible for the interaction with the interface to result in zero transmission. This phenomenon is called total internal reflection. Assuming that $n>n'$ and $\alpha=\arcsin\left(\frac{n'}{n}\right)$, then $\gamma=90°$, i.e., the total internal reflection of light will happen. Therefore, when the light travels from the optically thinner medium into the optically denser medium, the condition for total internal reflection can be expressed as

$$\alpha\geqslant\alpha_{\mathrm{cr}}=\arcsin\left(\frac{n'}{n}\right) \tag{1.3}$$

where α_{cr} is the critical angle. The total internal reflection allows for fiber optics technology. As light signals travel down a fiber optic cable, it undergoes total internal reflection allowing for essentially no light lost over the length of the cable. It is also possible to produce polarized light rays using a combination of reflection and refraction. When a refracted ray and the reflected ray form a right angle, the reflected ray has the property of plane polarization. The angle of incidence required for such a situation is known as Brewster's angle.

2. Lens

A device which produces converging or diverging light rays due to refraction is known as a lens. In general, two types of lenses exist: convex lenses, which cause parallel light rays to converge, and concave lenses, which cause parallel light rays to diverge. The detailed prediction of how images are produced by these lenses can be made using ray tracing similar to curved mirrors. Similarly to curved mirrors, thin lenses follow a simple equation that determines the location of the images according to the focal length (f) and object distance (a):

$$\frac{1}{a}+\frac{1}{b}=\frac{1}{f} \tag{1.4}$$

where b is the image distance and is considered by convention to be negative if on the same side of the lens as the object and positive if on the opposite side of the lens. The focal length f is considered negative for concave lenses.

Incoming parallel rays are focused by a convex lens into an inverted real image one focal length from the lens, on the far side of the lens. Rays from an object at finite distance are focused further from the lens than the focal distance; the closer the object is to the lens, the further the image is from the lens. With concave lenses, incoming parallel rays diverge after going through the lens, in such a way that they seem to have originated at an upright virtual image one focal length from the lens, on the same side of the lens that the parallel rays are approaching. Rays from an object at finite distance are associated with a virtual image that is closer to the lens than the focal length, and on the same side of the lens as the object. The closer the object is to the lens, the closer the virtual image is to the lens. The magnification of an imaging lens is given by

$$M=-\frac{b}{a} \tag{1.5}$$

where the negative sign is given, by convention, to indicate an upright image for positive values and an inverted image for negative values. Similar to mirrors, upright images produced by single lenses are virtual while inverted images are real.

Lenses suffer from aberrations that distort images. These are both due to geometrical imperfections and due to the changing index of refraction for different wavelengths of light (chromatic aberration).

1.1.2 Physical Optics

Physical optics, or wave optics, studies interference, diffraction, polarization, and other phenomena for which the light ray approximation of geometrical optics is not valid. Physical optics is a more comprehensive model of light, which includes wave effects such as interference and diffraction that cannot be accounted for in geometrical optics. Physical optics is also the name of an approximation, i.e., it is an intermediate method between geometrical optics, which ignores wave effects, and electromagnetism, which is a precise theory. This approximation consists of using geometrical optics to estimate the field on a lens, mirror or aperture and then integrating that field over the lens, mirror or aperture to calculate the transmitted or scattered field.

Light is an electromagnetic wave. Its direction of vibration is perpendicular to the propagation direction of light and light is therefore a transverse wave.

1. Wave Equation

Light satisfies the following wave equation:

$$\nabla^2 E(r,t) - \frac{1}{c^2}\frac{\partial^2 E(r,t)}{\partial t^2} = 0 \tag{1.6}$$

where $\nabla^2 = \frac{\partial^2}{\partial x^2} + \frac{\partial^2}{\partial y^2} + \frac{\partial^2}{\partial z^2}$ is the Laplace operator, $E(r,t)$ is the instantaneous light filed and c is the velocity of light. The monochromatic solution of the above wave equation can be expressed as

$$E(r,t) = A(r)\exp(-\mathrm{i}\omega t) \tag{1.7}$$

where $A(r)$ is the amplitude, $\mathrm{i} = \sqrt{-1}$ is the imaginary unit and ω is the circular frequency. Substituting Eq. (1.7) into Eq. (1.6), we obtain

$$(\nabla^2 + k^2)A(r) = 0 \tag{1.8}$$

where $k = \omega/c = 2\pi/\lambda$ is the wave number. Eq. (1.8) is called the Helmholtz equation.

2. Plane Wave

The plane wave solution of the wave equation can be written by

$$A(x,y,z) = a\exp(\mathrm{i}kz) \tag{1.9}$$

where a is the amplitude. Eq. (1.9) represents a plane wave which propagates along the z direction and has the intensity of the form

$$I(x,y,z) = |A(x,y,z)|^2 = a^2 \tag{1.10}$$

It is obvious that the intensity is the same at arbitrary point in a plane wave field.

3. Spherical Wave

The spherical wave solution of the wave equation can be expressed as

$$A(r) = \frac{a}{r}\exp(\mathrm{i}kr) \tag{1.11}$$

where a is the amplitude of the wave at unit distance from the light source. Eq. (1.11) represents a spherical wave which propagates outward and has the intensity

$$I(r) = |A(r)|^2 = \frac{a^2}{r^2} \tag{1.12}$$

i.e., the intensity at arbitrary point in a spherical wave field is inversely proportional to r^2.

A quadratic approximation to the spherical wave represented by Eq. (1.11)

$$A(x, y, z) = \frac{a}{z}\exp(ikz)\exp\left[\frac{ik}{2z}(x^2 + y^2)\right] \tag{1.13}$$

Eq. (1.13) represents the amplitude at any point (x, y, z) on a plane distant z from the point source.

4. Cylindrical Wave

The cylindrical wave solution of the wave equation can be given by

$$A(r) = \frac{a}{\sqrt{r}}\exp(ikr) \tag{1.14}$$

were a is the amplitude of the wave at unit distance. Eq. (1.14) represents a cylindrical wave which propagates outward with the intensity

$$I(r) = |A(r)|^2 = \frac{a^2}{r} \tag{1.15}$$

i.e., the intensity at arbitrary point in a cylindrical wave field is inversely proportional to r.

1.2 Optical Interference

Interference refers to the variation of resultant wave amplitude that occurs when two waves are superposed to form a resulting wave due to the interaction of waves that are coherent with each other. The principle of superposition of waves states that when two coherent waves are incident on the same point, the resultant amplitude at that point is equal to the vector sum of the amplitudes of the individual waves. If a crest of a wave meets a crest of another wave of the same frequency at the same point, then the amplitude is the sum of the individual amplitudes, i.e., constructive interference. If a crest of one wave meets a trough of another wave, then the amplitude is equal to the difference in the individual amplitudes, i.e., destructive interference. Constructive interference occurs when the phase difference between the waves is an even multiple of π, whereas destructive interference occurs when the difference is an odd multiple of π.

Because the frequency of light waves is too high to be detected by currently available detectors, it is possible to observe only the intensity of an optical interference pattern. The complex amplitude of the two waves at a point can be written as

$$A_1 = a_1\exp(i\varphi_1), \quad A_2 = a_2\exp(i\varphi_2) \tag{1.16}$$

where a_1 and a_2 are respectively the amplitudes of the two waves, and φ_1 and φ_2 are the phases of the two waves. The intensity of the light at a given point is proportional to the square of the amplitude of the wave; hence the intensity of the resulting wave can be expressed as

$$I = |A_1 + A_2|^2 = a_1^2 + a_2^2 + 2a_1a_2 \cos\Delta\varphi \tag{1.17}$$

where $\Delta\varphi = \varphi_2 - \varphi_1$ is the phase difference between the two waves. The equation above can also be expressed in terms of the intensities of the individual waves as

$$I = I_1 + I_2 + 2\sqrt{I_1I_2} \cos\Delta\varphi \tag{1.18}$$

where I_1 and I_2 are the intensities of the two waves, respectively. Thus it can be seen that the intensity distribution of the resulting wave contains cosine interference fringes. The contrast of fringes is defined as

$$V = \frac{I_{\max} - I_{\min}}{I_{\max} + I_{\min}} = \frac{2\sqrt{I_1I_2}}{I_1 + I_2} \tag{1.19}$$

where $I_{\max} = I_1 + I_2 + 2\sqrt{I_1I_2}$ and $I_{\min} = I_1 + I_2 - 2\sqrt{I_1I_2}$ are the maximum and minimum values of intensity, respectively. If the two beams are of equal intensity, the maxima are four times as bright as the individual beams, the minima have zero intensity, and then the contrast of fringes will reach a maximum value. Using Eq. (1.19), Eq. (1.18) can also be rewritten by

$$I = I_{\mathrm{B}} + I_{\mathrm{M}} \cos\varphi = I_{\mathrm{B}}(1 + V\cos\Delta\varphi) \tag{1.20}$$

where $I_{\mathrm{B}} = I_1 + I_2$ and $I_{\mathrm{M}} = 2\sqrt{I_1I_2}$ are respectively the background and modulation intensities.

The two waves must have the same frequency, the same polarization and an invariable difference in phase to give rise to interference fringes. The discussion above assumes that the waves which interfere with each other are monochromatic (i.e. a single frequency) and infinite in time. This is not, however, either practical or necessary. Two identical waves of finite duration whose frequency is fixed over that period will give rise to an interference pattern while they overlap. Two identical waves, each consists of a narrow frequency spectrum wave of finite duration, will give a series of fringe patterns of slightly differing spacings, and provided the spread of spacings is significantly less than the average fringe spacing, a fringe pattern will again be observed during the time when the two waves overlap.

1.3 Optical Diffraction

Diffraction refers to changes in the directions and intensities of a group of waves after passing through an aperture or by an obstacle. The diffraction phenomenon is described as the interference of waves according to the Huygens-Fresnel principle and exhibited when waves encounter an aperture or obstacle that is comparable in size to the wavelength of waves.

The effects of diffraction are often seen in everyday life. The most striking examples of diffraction are those that involve light; for example, the closely spaced tracks on a CD or DVD act as a diffraction grating to form the familiar rainbow pattern seen when looking at a disk. This principle can be extended to a grating with a structure such that it will produce any diffraction pattern desired; the hologram on a credit card is an example. Diffraction in the atmosphere by small particles can cause a bright ring to be visible around a bright light source like the sun or the moon. A shadow of a solid object, using light from a compact source, shows small fringes near its edges. The speckle pattern which is observed when laser light falls on an optically rough surface is also a diffraction phenomenon.

The propagation of a wave can be visualized by considering every particle of the transmitted medium on a wavefront as a point source for a secondary spherical wave. The wave amplitude at any subsequent point is the sum of these secondary waves. When waves are added together, their sum is determined by the relative phases as well as the amplitudes of the individual waves so that the summed amplitude of the waves can have any value between zero and the sum of the individual amplitudes. Hence, diffraction patterns usually have a series of maxima and minima.

There are various analytical models which allow the diffracted field to be calculated, including the Kirchhoff-Fresnel diffraction equation which is derived from wave equation, the Fresnel diffraction approximation which applies to the near field and the Fraunhofer diffraction approximation which applies to the far field. Most configurations cannot be solved analytically, but can yield numerical solutions through finite element and boundary element methods.

1. Fresnel Diffraction

Let $A(\xi,\eta)$, as shown in Fig. 1.2, be the amplitude distribution at the aperture plane (ξ,η) located at $z=0$. Under small-angle diffraction and near-axis approximation, the field at any point $P(x,y)$ at a plane distant z from the aperture plane can be expressed using Fresnel-Kirchhoff diffraction theory as follows:

$$A(x,y)=\frac{1}{\mathrm{i}\lambda z}\int_{-\infty}^{\infty}\int_{-\infty}^{\infty}A(\xi,\eta)\exp\left[\mathrm{i}k\sqrt{z^2+(x-\xi)^2+(y-\eta)^2}\right]\mathrm{d}\xi\mathrm{d}\eta \tag{1.21}$$

where the integral is over the area of the aperture and $k=2\pi/\lambda$ is the wave number of light. In the Fresnel diffraction region, using the following condition:

$$z^3 \gg \frac{\pi}{4\lambda}[(x-\xi)^2+(y-\eta)^2]^2_{\max} \tag{1.22}$$

then the amplitude $A(x,y)$ can be given by

$$A(x,y)=\frac{\exp(\mathrm{i}kz)}{\mathrm{i}\lambda z}\exp\left[\mathrm{i}\frac{\pi}{\lambda z}(x^2+y^2)\right]\int_{-\infty}^{\infty}\int_{-\infty}^{\infty}A(\xi,\eta)\exp\left[\mathrm{i}\frac{\pi}{\lambda z}(\xi^2+\eta^2)\right]\exp\left[-\mathrm{i}\frac{2\pi}{\lambda z}(x\xi+y\eta)\right]\mathrm{d}\xi\mathrm{d}\eta \tag{1.23}$$

Thus, aside from multiplicative amplitude and phase factors that are independent of (ξ,η), the function $A(x,y)$ can be found from a Fourier transform of $A(\xi,\eta)\exp\left[\mathrm{i}\frac{\pi}{\lambda z}(\xi^2+\eta^2)\right]$.

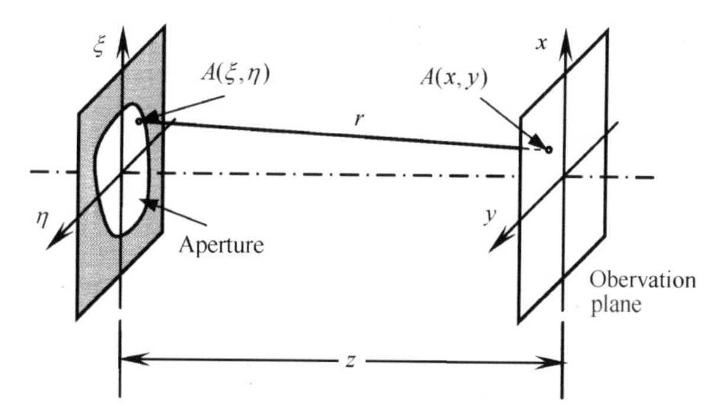

Fig. 1.2 Diffraction

2. Fraunhofer Diffraction

If the far-field condition

$$z \gg \frac{\pi}{\lambda}(\xi^2+\eta^2)_{\max} \tag{1.24}$$

is met we are in the Fraunhofer diffraction region. The amplitude $A(x,y)$ in the Fraunhofer diffraction region can be written as

$$A(x,y)=\frac{\exp(\mathrm{i}kz)}{\mathrm{i}\lambda z}\exp\left[\mathrm{i}\frac{\pi}{\lambda z}(x^2+y^2)\right]\int_{-\infty}^{\infty}\int_{-\infty}^{\infty}A(\xi,\eta)\exp\left[-\mathrm{i}\frac{2\pi}{\lambda z}(x\xi+y\eta)\right]\mathrm{d}\xi\mathrm{d}\eta \tag{1.25}$$

It can be seen that, aside from the multiplicative factors located in front of the integral, $A(x,y)$ can be expressed as the Fourier transform of the aperture function.

1.4 Optical Polarization

Polarization refers to a state in which light rays exhibit different properties in different directions, especially the state in which all the vibrations take place in one plane.

In an electromagnetic wave, both the electric field and magnetic field are oscillating but in different directions; by convention the polarization of light refers to the polarization of the electric field. The oscillation of the electric field may be in a single direction (i.e. linear polarization), or the field may rotate at the optical frequency (i.e. circular or elliptical polarization).

Most sources of light, including thermal (black body) radiation and fluorescence (but not lasers), are classified as incoherent and unpolarized (or only partially polarized) because they consist of a random mixture of waves having different spatial characteristics, frequencies (wavelengths), phases, and polarization states. These sources of light produce light described as incoherent. Radiation is produced independently by a large number of atoms or molecules whose emissions are uncorrelated and generally of random polarizations. In this case the light is said to be unpolarized.

Light is said to be partially polarized when there is more power in one polarization mode than another. At any particular wavelength, partially polarized light can be statistically described as the superposition of a completely unpolarized component, and a completely polarized one. One may then describe the light in terms of the degree of polarization, and the parameters of the polarized component.

The most common optical materials (such as glass) are isotropic and simply preserve the polarization of a wave but do not differentiate between polarization states. However there are important classes of materials classified as birefringent in which this is not the case and a wave's polarization will generally be modified or will affect propagation through it. A polarizer is an optical filter that transmits only one polarization.

1.5 Optical Interferometers

Interferometry is a family of techniques in which waves, usually electromagnetic, are superimposed in order to extract phase information about the waves. An interferometer is a device that uses interference phenomena between a reference wave and an object wave or between two parts of an object wave to determine displacements, velocities, etc. Typically (for example, a Michelson interferometer) a single incoming beam of coherent light will be split into two identical beams by a beamsplitter (a partially reflecting mirror). Each of these beams travels a different route or path, and they are recombined when arriving at a detector. The path difference, the difference in the distance traveled by each beam, creates a phase difference between them. It is this introduced phase difference that creates the interference pattern between the two identical beams.

Traditionally, interferometers have been classified as either amplitude-division or wavefront-division systems. In an amplitude-division system, a beam splitter is used to divide the light into two beams traveling in different directions, which are then superimposed to produce the interference pattern. The Michelson interferometer and the Mach-Zehnder interferometer are examples of amplitude-division systems. In a wavefront-division system, the wave is divided in space. For example, Young's double slit interferometer and Lloyd's mirror.

1. Michelson Interferometer

The Michelson interferometer is common configuration for optical interferometry. Using a beamsplitter, a light source is split into two arms. Each of those is reflected back toward the beamsplitter which then combines coherently their amplitudes. Depending on the interferometer's particular application, the two paths may be of different lengths or include optical materials or components under test.

A Michelson interferometer consists minimally of two mirrors and a beamsplitter. In Fig. 1.3, a laser source emits light that hits the surface of a plate beamsplitter. The beamsplitter is partially reflective, so part of the light is transmitted through to the moving mirror while some is reflected in the direction of the fixed mirror. Both beams recombine on the beamsplitter to produce an interference pattern incident on the detector (or on the retina of a person's eye). If there is a slight angle between the two returning beams, for instance, then an imaging detector will record a sinusoidal fringe pattern. If there is perfect spatial alignment between the returning beams, then there will not be any such pattern but rather a constant intensity over the beam dependent on the optical path difference.

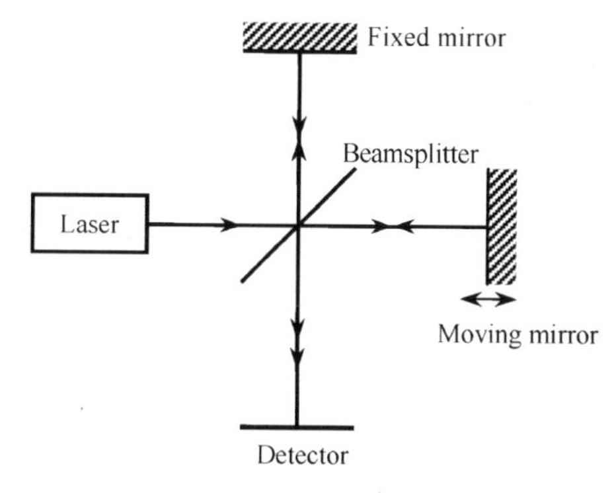

Fig. 1.3 Michelson interferometer

The extent of the fringes depends on the coherence length of the source. Single longitudinal mode lasers are highly coherent and can produce high contrast interference with optical path difference of millions or even billions of wavelengths.

The Twyman-Green interferometer is a variation of the Michelson interferometer used to test small optical components. The main characteristics distinguishing it from the Michelson configuration are the use of a monochromatic point light source and a collimator.

2. Mach-Zehnder Interferometer

The Mach-Zehnder interferometer is a device used to determine the relative phase variations between two collimated beams derived by splitting light from a single source. The interferometer has been used to measure phase shifts between the two beams caused by a sample or a change in length of one of the paths.

The Mach-Zehnder interferometer is a highly configurable instrument, as shown in Fig. 1.4. In contrast to the well-known Michelson interferometer, each of the well separated light paths is traversed only once. A collimated beam is split by a half-silvered mirror. The two beams (i.e. the sample beam and the reference beam) are each reflected by a mirror. The two beams then pass a second half-silvered mirror and enter a detector.

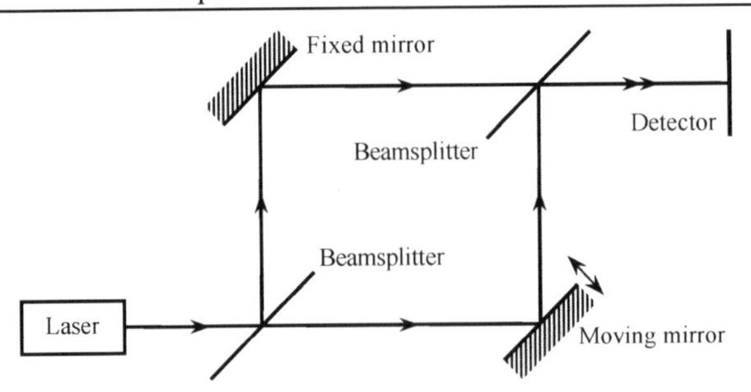

Fig. 1.4 Mach-Zehnder interferometer

Collimated sources result in a nonlocalized fringe pattern. In most cases, the fringes would be adjusted to lie in the same plane as the test object, so that fringes and test object can be photographed together.

3. Sagnac Interferometer

The Sagnac effect (also called Sagnac interference) is a phenomenon encountered in interferometry that is elicited by rotation. The Sagnac effect manifests itself in a setup called a ring interferometer. A beam of light is split and the two beams are made to follow the same path but in opposite directions. To act as a ring the trajectory must enclose an area. On return to the point of entry the two light beams are allowed to exit the ring and undergo interference. The relative phases of the two exiting beams, and thus the position of the interference fringes, are shifted according to the angular velocity of the apparatus. This arrangement is also called a Sagnac interferometer.

Typically three mirrors are used in a Sagnac interferometer, so that counter-propagating light beams follow a closed path such as a square as shown in Fig. 1.5. If the platform on which the ring interferometer is mounted is rotating, the interference fringes are displaced compared to their position when the platform is not rotating. The amount of displacement is proportional to the angular velocity of the rotating platform. The axis of rotation does not have to be inside the enclosed area.

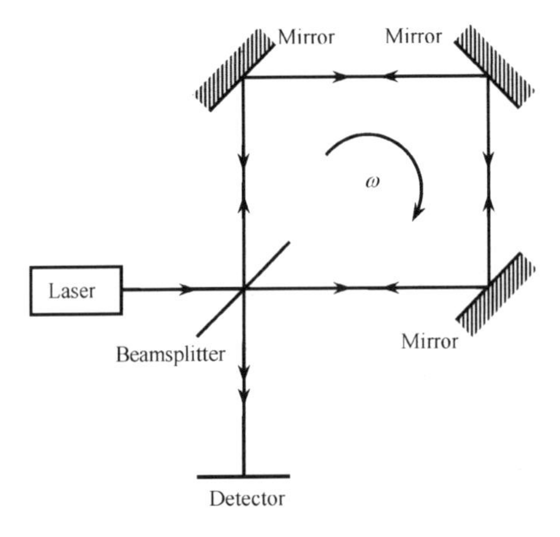

Fig. 1.5 Sagnac interferometer

4. Fabry-Pérot Interferometer

A Fabry-Pérot interferometer is typically made of a transparent plate with two reflecting surfaces, or two parallel highly reflecting mirrors, as shown in Fig. 1.6. The heart of the Fabry-Pérot interferometer is a pair of partially reflective optical flats spaced micrometers to centimeters apart, with the reflective surfaces facing each other. The flats in an interferometer are often made in a wedge shape to prevent the rear surfaces from producing interference fringes; the rear surfaces often also have an antireflective coating.

In a typical system, illumination is provided by a diffuse source located at the focal plane of a collimating lens. A focusing lens after the pair of flats would produce an inverted image of the source if the flats were not present; all light emitted from a point on the source is focused to a single point in the image plane of the system. As the ray passes through the paired flats, it is multiply reflected to produce multiple transmitted rays which are collected by the focusing lens and brought to a point on the screen. The complete interference pattern takes the appearance of a set of concentric rings.

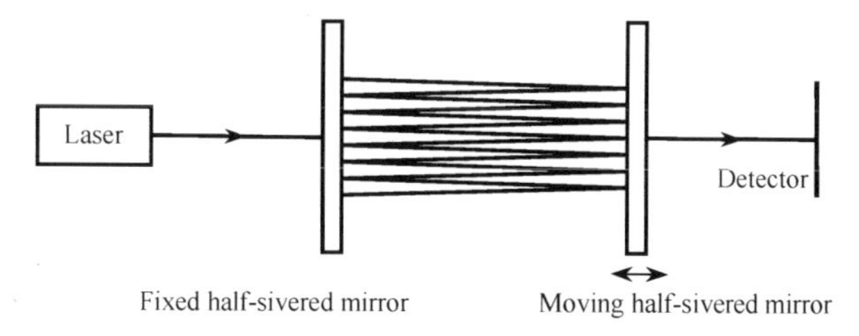

Fig. 1.6 Fabry-Pérot interferometer

1.6 Lasers

A laser is a device that converts incident light (electromagnetic radiation) of mixed frequencies to one or more discrete frequencies of highly amplified and coherent infrared, visible, or ultraviolet light. The term "laser" is an acronym for "Light Amplification by Stimulated Emission of Radiation".

A laser differs from other sources of light by its coherence. Spatial coherence allows a laser to be focused to a tight spot, achieving a very high irradiance. Spatial coherence also allows a laser beam to stay narrow over great distances (collimation). Lasers can also have high temporal (or longitudinal) coherence, which allows them to emit light with a very narrow spectrum; i.e., they can emit a single color of light. Temporal coherence implies a polarized wave at a single frequency whose phase is correlated over a relatively great distance (the coherence length) along the beam.

Most lasers actually produce radiation in several modes having slightly differing frequencies (wavelengths), often not in a single polarization. Although temporal coherence implies monochromaticity, there are lasers that emit a broad spectrum of light or emit different wavelengths of light simultaneously. All such devices are also classified as lasers based on their method of producing light, i.e., stimulated emission.

Lasers are employed in applications where light of the required spatial or temporal coherence could not be produced. Lasers can be used in optical disk drives, laser printers, and barcode scanners; fiber-optic and free-space optical communication; laser surgery and skin treatments; cutting and welding materials.

A beam produced by a thermal or other incoherent light source has an instantaneous amplitude and phase that vary randomly with respect to time and position, thus having a short coherence length.

Chapter 2 Holography and Holographic Interferometry

Holography has not received widespread attention in the next more than ten years since it was proposed for the first time, because highly coherent light sources were not available and two twin images could not be separated. It was not until inventing the laser and proposing the off-axis method that holography started to have a rapid development. Since then, various holographic methods (e.g., double-exposure holographic interferometry, time-averaged holographic interferometry, etc.) have been proposed, and various holographic applications (e.g., displacement measurement, vibration analysis, etc.) have been emerging.

2.1 Holography

Holography refers to a method used for producing three-dimensional images of objects; i.e., holography can be used for both recording the complex amplitude of an object wave, rather than the intensity as is the case in photography, on a photographic plate by interference between the object and reference waves and then reconstructing the complex amplitude of the object wave by diffraction of the hologram recorded on the photographic plate when illuminated by a reconstruction wave.

Holography was proposed by the Hungarian-British physicist Dennis Gabor for the first time. Therefore, he was awarded the Nobel Prize in Physics in 1971 for his invention of holography in 1948. Gabor recorded a hologram by illuminating a photographic plate with two in-line light waves; thus this method is usually called in-line holography. Because the object and reference waves are parallel in in-line holography, an in-line hologram will result in the virtual image superposed by the real image and the undiffracted reconstruction wave. A significant improvement of in-line holography was made by Emmett Leith et al., who introduced an off-axis reference wave, called off-axis holography. In off-axis holography, the virtual image, the real image, and the undiffracted reconstruction wave can be separated spatially.

2.1.1 Holographic Recording

The recording system used in holography is shown in Fig. 2.1. Laser with sufficient coherence is split into two waves by a beamsplitter placed in the system. The first wave, called the object wave, is used to illuminates the object. It is then scattered at the object surface and reflected to the photographic plate. The second wave, called the reference wave, illuminates the photographic plate directly. These two waves will interfere with each other to form an interference pattern (called hologram) on the photographic plate.

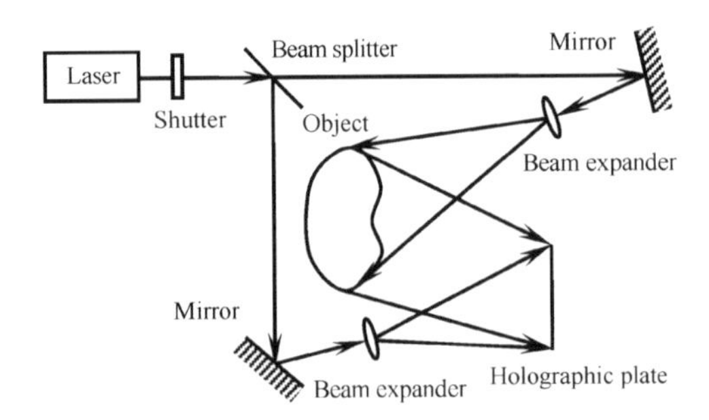

Fig.2.1 Recording system

Assuming that the complex amplitudes of the object and reference waves on the photographic plate are, respectively, denoted by $O = a_{\mathrm{o}} \exp(\mathrm{i}\varphi_{\mathrm{o}})$ and $R = a_{\mathrm{r}} \exp(\mathrm{i}\varphi_{\mathrm{r}})$ with a_{o} and φ_{o} being the amplitude and phase of the object wave, a_{r} and φ_{r} being the amplitude and phase of the reference wave, and $\mathrm{i} = \sqrt{-1}$ is the imaginary unit, then the intensity distribution recorded on the photographic plate can be expressed as

$$I = (O + R)(O + R)^* = (a_{\mathrm{o}}^2 + a_{\mathrm{r}}^2) + a_{\mathrm{o}} a_{\mathrm{r}} \exp[\mathrm{i}(\varphi_{\mathrm{o}} - \varphi_{\mathrm{r}})] + a_{\mathrm{o}} a_{\mathrm{r}} \exp[-\mathrm{i}(\varphi_{\mathrm{o}} - \varphi_{\mathrm{r}})] \tag{2.1}$$

where $*$ denotes the complex conjugate. Using $\cos\theta = [\exp(\mathrm{i}\theta) + \exp(-\mathrm{i}\theta)]/2$, the above equation can be rewritten by

$$I = (a_{\mathrm{o}}^2 + a_{\mathrm{r}}^2) + 2a_{\mathrm{o}} a_{\mathrm{r}} \cos(\varphi_{\mathrm{o}} - \varphi_{\mathrm{r}}) \tag{2.2}$$

Assuming that the exposure time is equal to T, the exposure of the photographic plate can then be given by

$$E = IT = T(a_{\mathrm{o}}^2 + a_{\mathrm{r}}^2) + Ta_{\mathrm{o}} a_{\mathrm{r}} \exp[\mathrm{i}(\varphi_{\mathrm{o}} - \varphi_{\mathrm{r}})] + Ta_{\mathrm{o}} a_{\mathrm{r}} \exp[-\mathrm{i}(\varphi_{\mathrm{o}} - \varphi_{\mathrm{r}})] \tag{2.3}$$

Assuming that the amplitude transmittance is proportional to the exposure, then the amplitude transmittance of the hologram can be expressed as

$$t = \beta E = \beta T(a_{\mathrm{o}}^2 + a_{\mathrm{r}}^2) + \beta Ta_{\mathrm{o}} a_{\mathrm{r}} \exp[\mathrm{i}(\varphi_{\mathrm{o}} - \varphi_{\mathrm{r}})] + \beta Ta_{\mathrm{o}} a_{\mathrm{r}} \exp[-\mathrm{i}(\varphi_{\mathrm{o}} - \varphi_{\mathrm{r}})] \tag{2.4}$$

where β is a constant of proportionality.

2.1.2 Holographic Reconstruction

The original object wave can be reconstructed by illuminating the hologram with the original reference wave, as shown in Fig. 2.2. An observer can see a virtual image, which is indistinguishable from the original object.

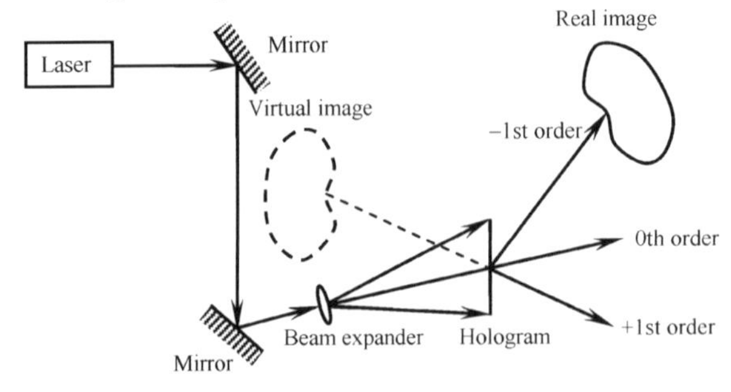

Fig. 2.2 Reconstruction system

When the hologram is illuminated with the reference wave $R = a_r \exp(i\varphi_r)$, the complex amplitude of the light wave passing through the hologram can be described by

$$A = Rt = \beta T(a_o^2 a_r + a_r^3)\exp(i\varphi_r) + \beta T a_o a_r^2 \exp(i\varphi_o) + \beta T a_o a_r^2 \exp[-i(\varphi_o - 2\varphi_r)] \tag{2.5}$$

where $\beta T(a_o^2 a_r + a_r^3)\exp(i\varphi_r)$ is the zeroth order diffraction wave, $\beta T a_o a_r^2 \exp(i\varphi_o)$ is the positive first order diffraction wave and represents the virtual image of the object, and $\beta T a_o a_r^2 \exp[-i(\varphi_o - 2\varphi_r)]$ is the negative first order diffraction wave and represents the real image (i.e., conjugate image) of the object.

2.2 Holographic Interferometry

A major application of holography is holographic interferometry. Holographic interferometry is a high-sensitivity, non-contact, and full-field method. It can provide comparison of two different states of the object with optically rough surface and enable the object displacement to be measured to optical interferometric precision, i.e. to fractions of a wavelength of light. Therefore, holographic interferometry can be applied to deformation measurement, vibration analysis, and non-destructive test. It can also be used to generate contour fringes representing the surface shape or form of the object to be measured.

2.2.1 Phase Calculation

Assume that an object point is moved from P to P' after the object is deformed and that the displacement of this object point is denoted by $\boldsymbol{d}$, as shown in Fig. 2.3. S and O denote the positions of the light source and observation point, respectively. The position vectors of P and P' with respect to S are respectively $\boldsymbol{r}_0$ and $\boldsymbol{r}_0'$, and the position vectors of O relative to P and P' are respectively $\boldsymbol{r}$ and $\boldsymbol{r}'$. Therefore, when the object point is moved from P to P', the phase change of light wave can be expressed as

$$\delta = k[(\boldsymbol{e}_0 \cdot \boldsymbol{r}_0 + \boldsymbol{e} \cdot \boldsymbol{r}) - (\boldsymbol{e}_0' \cdot \boldsymbol{r}_0' + \boldsymbol{e}' \cdot \boldsymbol{r}')] \tag{2.6}$$

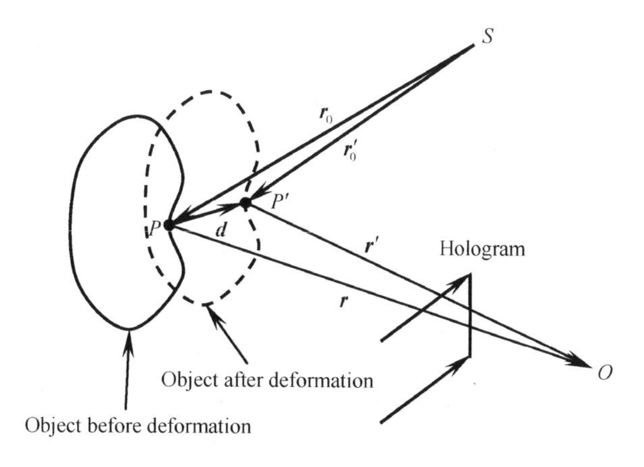

Fig. 2.3 Phase calculation

where $k = 2\pi/\lambda$ is the wave number of the light wave, with λ being the wavelength, $\boldsymbol{e}_0$ and $\boldsymbol{e}'_0$ are the unit vectors along $\boldsymbol{r}_0$ and $\boldsymbol{r}'_0$ respectively, $\boldsymbol{e}$ and $\boldsymbol{e}'$ are respectively the unit vectors along $\boldsymbol{r}$ and $\boldsymbol{r}'$.

For small deformation, we have $\boldsymbol{e}'_0 \approx \boldsymbol{e}_0$, $\boldsymbol{e}' \approx \boldsymbol{e}$. Hence Eq. (2.6) can be rewritten by

$$\delta = k[\boldsymbol{e}_0 \cdot (\boldsymbol{r}_0 - \boldsymbol{r}'_0) + \boldsymbol{e} \cdot (\boldsymbol{r} - \boldsymbol{r}')] \tag{2.7}$$

Using $\boldsymbol{r}_0 - \boldsymbol{r}'_0 = -\boldsymbol{d}$, $\boldsymbol{r} - \boldsymbol{r}' = \boldsymbol{d}$, we obtain

$$\delta = k(\boldsymbol{e} - \boldsymbol{e}_0) \cdot \boldsymbol{d} \tag{2.8}$$

2.2.2 Double-Exposure Holographic Interferometry

In double-exposure holographic interferometry, two different states of a deformed object are recorded on a single photographic plate through two series of exposure. The first exposure represents the object in its reference state; the second exposure represents the object in its loaded state. When the photographic plate after development and fixation is placed into its original position to be reconstructed by the reference wave, two reconstructed object waves, which correspond to two different states of the object, will interfere with each other to produce interference fringes related to the object surface displacement. Because two recordings have slightly different object waves, only one image superposed by interference fringes is visible. Therefore, the object surface displacement can be determined through analyzing these interference fringes located on or close to the object surface.

The recording system used in double-exposure holographic interferometry is shown in Fig. 2.4. Assuming that the complex amplitudes of the object waves before and after deformation are respectively denoted by $O_1 = a_\text{o} \exp(\text{i}\varphi_\text{o})$ and $O_2 = a_\text{o} \exp[\text{i}(\varphi_\text{o} + \delta)]$, where a_o being the amplitude of the object wave (assume that the amplitude of the object wave is unchanged during deformation), φ_o the phase of the object wave before deformation, and δ the phase change of the object wave caused due to the object deformation, and that the complex amplitude of the reference wave is equal to $R = a_\text{r} \exp(\text{i}\varphi_\text{r})$, then the intensity distribution recorded on the photographic plate when subjected to two series of exposure can given, respectively, by

$$\begin{aligned} I_1 &= (O_1 + R) \cdot (O_1 + R)^* = (a_\text{o}^2 + a_\text{r}^2) + a_\text{o} a_\text{r} \exp[\text{i}(\varphi_\text{o} - \varphi_\text{r})] + a_\text{o} a_\text{r} \exp[-\text{i}(\varphi_\text{o} - \varphi_\text{r})] \\ I_2 &= (O_2 + R) \cdot (O_2 + R)^* = (a_\text{o}^2 + a_\text{r}^2) + a_\text{o} a_\text{r} \exp[\text{i}(\varphi_\text{o} + \delta - \varphi_\text{r})] + a_\text{o} a_\text{r} \exp[-\text{i}(\varphi_\text{o} + \delta - \varphi_\text{r})] \end{aligned} \tag{2.9}$$

If the exposure time is equal to T for each exposure, the total exposure recorded on the photographic plate is equal to

$$\begin{aligned} E = (I_1 + I_2)T = 2(a_\text{o}^2 + a_\text{r}^2)T &+ a_\text{o} a_\text{r} T \exp[\text{i}(\varphi_\text{o} - \varphi_\text{r})][1 + \exp(\text{i}\delta)] \\ &+ a_\text{o} a_\text{r} T \exp[-\text{i}(\varphi_\text{o} - \varphi_\text{r})][1 + \exp(-\text{i}\delta)] \end{aligned} \tag{2.10}$$

Assuming that the amplitude transmittance is proportional to the exposure, then the amplitude transmittance of the hologram can be written by

$$t = \beta E = 2\beta(a_o^2 + a_r^2)T + \beta a_o a_r T \exp[i(\varphi_o - \varphi_r)][1 + \exp(i\delta)] + \beta a_o a_r T \exp[-i(\varphi_o - \varphi_r)][1 + \exp(-i\delta)] \tag{2.11}$$

where β is a constant of proportionality.

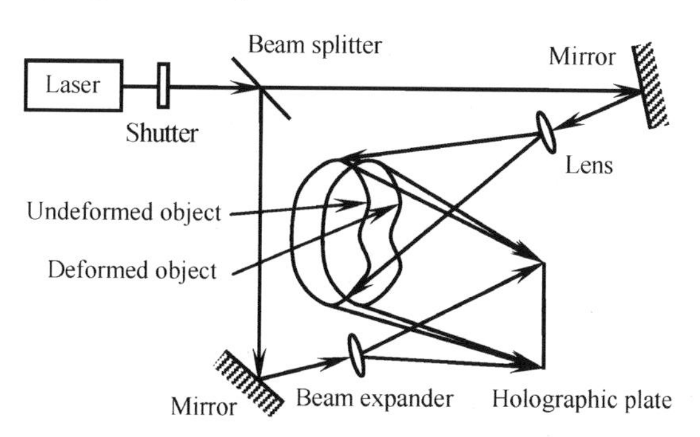

Fig. 2.4 Double-exposure recording system

The reconstruction system for double-exposure holographic interferometry is shown in Fig. 2.5. When the hologram is illuminated with the original reference wave $R = a_r \exp(i\varphi_r)$, the complex amplitude of the light wave passing through the hologram can be described by

$$A = Rt = 2\beta(a_o^2 a_r + a_r^3)T \exp(i\varphi_r) + \beta a_o a_r^2 T \exp(i\varphi_o)[1 + \exp(i\delta)] + \beta a_o a_r^2 T \exp[-i(\varphi_o - 2\varphi_r)][1 + \exp(-i\delta)] \tag{2.12}$$

where the first term is the zeroth order diffraction wave, the second term the positive first order diffraction wave, and the third term the negative first order diffraction wave.

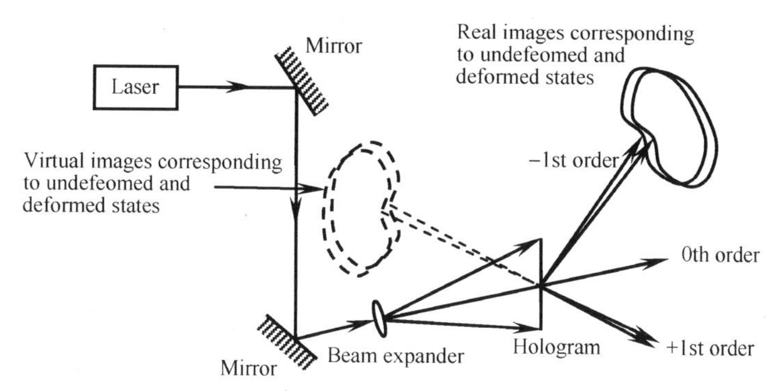

Fig. 2.5 Double-exposure holographic reconstruction

If only the positive first order diffraction wave is considered, the complex amplitude passing through the hologram can be expressed as

$$A' = \beta a_o a_r^2 T \exp(i\varphi_o)[1 + \exp(i\delta)] \tag{2.13}$$

The corresponding intensity distribution is equal to

$$I' = A'A'^* = 2(\beta a_o a_r^2 T)^2[1 + \cos\delta] \tag{2.14}$$

It is obvious that when the condition

$$\delta = 2n\pi \quad (n = 0, \pm 1, \pm 2, \cdots) \tag{2.15}$$

is satisfied, bright fringes will be formed, and that when

$$\delta = (2n+1)\pi \quad (n = 0, \pm 1, \pm 2, \cdots) \tag{2.16}$$

is satisfied, dark fringes will be formed.

Using $\delta = k(\boldsymbol{e} - \boldsymbol{e}_0) \cdot \boldsymbol{d}$, we obtain

$$I' = 2(\beta T a_o a_r^2)^2 \{1 + \cos[k(\boldsymbol{e} - \boldsymbol{e}_0) \cdot \boldsymbol{d}]\} \tag{2.17}$$

Thus bright fringes will be formed when

$$(\boldsymbol{e} - \boldsymbol{e}_0) \cdot \boldsymbol{d} = n\lambda \quad (n = 0, \pm 1, \pm 2, \cdots) \tag{2.18}$$

and dark fringes will be formed when

$$(\boldsymbol{e} - \boldsymbol{e}_0) \cdot \boldsymbol{d} = \left(n + \frac{1}{2}\right)\lambda \quad (n = 0, \pm 1, \pm 2, \cdots) \tag{2.19}$$

2.2.3 Real-Time Holographic Interferometry

Holography enables the object wave to be recorded on a photographic plate and to be reconstructed by illuminating this photographic plate with the original reference wave. If the reconstructed object wave is superposed on the actual object wave scattered from the same object, then these two object waves will be exactly the same. If, however, a small deformation is applied to the object, the relative phase between these two object waves will alter, and it is possible to observe interference fringes in real time. This method is known as real-time holographic interferometry. In this method, the hologram after wet processing is required to be replaced exactly in the original recording position so that when it is illuminated with the original reference wave, the reconstructed virtual image will coincide exactly with the object.

Assuming that the complex amplitudes of the object and reference waves before deformation are, respectively, denoted by $O = a_o \exp(\mathrm{i}\varphi_o)$ and $R = a_r \exp(\mathrm{i}\varphi_r)$ with a_o and φ_o being the amplitude and phase of the object wave, and a_r and φ_r being the amplitude and phase of the reference wave, then the intensity distribution before deformation recorded on a photographic plate can be expressed as

$$I = (O+R)(O+R)^* = (a_o^2 + a_r^2) + a_o a_r \exp[\mathrm{i}(\varphi_o - \varphi_r)] + a_o a_r \exp[-\mathrm{i}(\varphi_o - \varphi_r)] \tag{2.20}$$

Assuming that the exposure time is equal to T, that the amplitude transmittance is proportional to the exposure, and that the constant of proportionality is denoted by β, then the amplitude transmittance of the hologram after development and fixation can be written by

$$t = \beta I T = \beta T(a_o^2 + a_r^2) + \beta T a_o a_r \exp[\mathrm{i}(\varphi_o - \varphi_r)] + \beta T a_o a_r \exp[-\mathrm{i}(\varphi_o - \varphi_r)] \tag{2.21}$$

The above hologram is replaced exactly in the original recording position and illuminated simultaneously with the original object and reference waves, as shown in Fig. 2.6. If the object is now subjected to deformation, then the complex amplitude of the object wave after deformation can be given by $O' = a_o \exp[i(\varphi_o + \delta)]$, where δ being the phase change of the object wave due to the object deformation. Therefore, the complex amplitude of the object wave passing through the hologram can be expressed as

$$\begin{aligned} A = (O' + R)t = \beta T(a_o^3 + a_o a_r^2)\exp[i(\varphi_o + \delta)] + \beta T a_o^2 a_r \exp[i(2\varphi_o - \varphi_r + \delta)] \\ + \beta T a_o^2 a_r \exp[i(\varphi_r + \delta)] + \beta T(a_o^2 a_r + a_r^3)\exp(i\varphi_r) \\ + \beta T a_o a_r^2 \exp(i\varphi_o) + \beta T a_o a_r^2 \exp[-i(\varphi_o - 2\varphi_r)] \end{aligned} \quad (2.22)$$

where the first term $\beta T(a_o^3 + a_o a_r^2)\exp[i(\varphi_o + \delta)]$ and the fifth term $\beta T a_o a_r^2 \exp(i\varphi_o)$ are related to the object wave. If only these two terms are considered, then we obtain

$$A' = \beta T \exp(i\varphi_o)[a_o a_r^2 + (a_o^3 + a_o a_r^2)\exp(i\delta)] \quad (2.23)$$

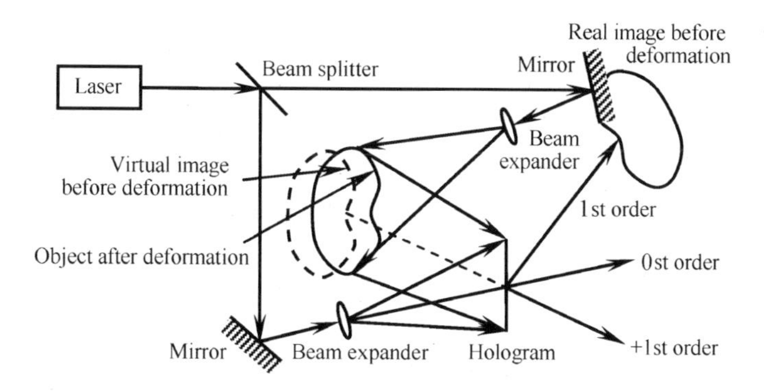

Fig. 2.6 Real-time reconstruction system

Using $I_o \ll I_r$, i.e. $a_o^2 \ll a_r^2$, we have

$$A' = \beta T a_o a_r^2 \exp(i\varphi_o)[1 + \exp(i\delta)] \quad (2.24)$$

And the corresponding intensity distribution is equal to

$$I' = A'A'^* = 2(\beta T a_o a_r^2)^2(1 + \cos\delta) \quad (2.25)$$

By substitution of $\delta = k(\boldsymbol{e} - \boldsymbol{e}_0) \cdot \boldsymbol{d}$, we obtain

$$I' = 2(\beta T a_o a_r^2)^2\{1 + \cos[k(\boldsymbol{e} - \boldsymbol{e}_0) \cdot \boldsymbol{d}]\} \quad (2.26)$$

Therefore, when the condition

$$(\boldsymbol{e} - \boldsymbol{e}_0) \cdot \boldsymbol{d} = n\lambda \quad (n = 0, \pm 1, \pm 2, \cdots) \quad (2.27)$$

is satisfied, bright fringes will be formed; and when

$$(\boldsymbol{e} - \boldsymbol{e}_0) \cdot \boldsymbol{d} = \left(n + \frac{1}{2}\right)\lambda \quad (n = 0, \pm 1, \pm 2, \cdots) \quad (2.28)$$

is met, dark fringes will also be formed.

2.2.4 Time-Averaged Holographic Interferometry

Time-averaged holographic interferometry is usually used for vibration analysis. As the name suggests, the recording is carried out over a time duration that is several times the period of vibration. This method will record all the states of vibration on a single photographic plate through continuous exposure. When the photographic plate after development and fixation is placed into its original position to be reconstructed by the original reference wave, all the reconstructed object waves corresponding to all the states will interfere with each other to form interference fringes related to the amplitude distribution of the vibrating object.

Assuming that the complex amplitude of a vibrating object at any time is denoted by $O = a_o \exp[i(\varphi_o + \delta)]$, where a_o is the amplitude of the object wave, $(\varphi_o + \delta)$ is the phase of object wave, and δ is the phase change of the object wave caused due to vibration. Assuming that the object is in simple harmonic vibration, then the equation of vibration can be expressed as

$$\boldsymbol{d} = \boldsymbol{A} \sin \omega t \tag{2.29}$$

where $\boldsymbol{A}$ is the amplitude vector of vibration, and ω is the circular or angular frequency of vibrating object. Therefore the phase change of the object wave at any time is equal to

$$\delta = k(\boldsymbol{e} - \boldsymbol{e}_0) \cdot \boldsymbol{d} = k(\boldsymbol{e} - \boldsymbol{e}_0) \cdot \boldsymbol{A} \sin \omega t \tag{2.30}$$

Assuming that the complex amplitude of the reference wave is $R = a_r \exp(i\varphi_r)$, then when the object is vibrating, the intensity distribution at any time recorded on a photographic plate can be given by

$$I = (O + R)(O + R)^* = (a_o^2 + a_r^2) + a_o a_r \exp[i(\varphi_o - \varphi_r + \delta)] + a_o a_r \exp[-i(\varphi_o - \varphi_r + \delta)] \tag{2.31}$$

Assuming that the exposure time is equal to T, and that the second term related to the object wave is considered, the total exposure recorded on the photographic plate is equal to

$$E = \int_0^T a_o a_r \exp[i(\varphi_o - \varphi_r + \delta)] dt = a_o a_r \exp[i(\varphi_o - \varphi_r)] \int_0^T \exp[ik(\boldsymbol{e} - \boldsymbol{e}_0) \cdot \boldsymbol{A} \sin \omega t] dt \tag{2.32}$$

Using $T \gg 2\pi/\omega$, the above equation can be rewritten

$$E = T a_o a_r \exp[i(\varphi_o - \varphi_r)] J_0[k(\boldsymbol{e} - \boldsymbol{e}_0) \cdot \boldsymbol{A}] \tag{2.33}$$

where J_0 is the zeroth order Bessel function of the first kind.

Assuming that the amplitude transmittance is proportional to the exposure, and that the constant of proportionality is denoted by β, then the amplitude transmittance of the hologram after development and fixation can be expressed as

$$t = \beta E = \beta T a_o a_r \exp[i(\varphi_o - \varphi_r)] J_0[k(\boldsymbol{e} - \boldsymbol{e}_0) \cdot \boldsymbol{A}] \tag{2.34}$$

When the hologram is illuminated by the original reference wave $R = a_r \exp(i\varphi_r)$, the complex amplitude of the light wave passing through the hologram can be written by

$$A' = Rt = \beta T a_o a_r^2 \exp(i\varphi_o) J_0[k(\boldsymbol{e} - \boldsymbol{e}_0) \cdot \boldsymbol{A}] \tag{2.35}$$

and the corresponding intensity distribution is equal to

$$I' = A'A'^* = (\beta T a_o a_r^2)^2 J_0^2[k(\boldsymbol{e}-\boldsymbol{e}_0)\cdot \boldsymbol{A}] \tag{2.36}$$

Eq. (2.36) shows that the intensity distribution is related to $J_0^2[k(\boldsymbol{e}-\boldsymbol{e}_0)\cdot \boldsymbol{A}]$. When $J_0^2[k(\boldsymbol{e}-\boldsymbol{e}_0)\cdot \boldsymbol{A}]$ is maximum, bright fringes will be formed. Dark fringes will be formed when $J_0^2[k(\boldsymbol{e}-\boldsymbol{e}_0)\cdot \boldsymbol{A}]$ is minimum; i.e.

$$k(\boldsymbol{e}-\boldsymbol{e}_0)\cdot \boldsymbol{A} = \alpha \quad (\alpha = 2.41,\ 5.52,\ 8.65,\ 11.79,\ 14.98,\ \cdots) \tag{2.37}$$

where α is the root of the zeroth order Bessel function; i.e., $J_0(\alpha)=0$. Therefore, the amplitude distribution of vibration can be determined when the orders of fringes are determined. The $J_0^2(\alpha)$-α curve is shown in Fig. 2.7.

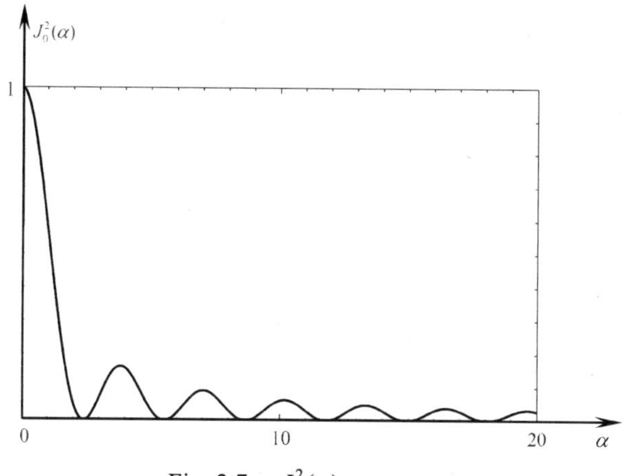

Fig. 2.7 $J_0^2(\alpha)$-α curve

2.2.5 Real-Time Time-Averaged Holographic Interferometry

In real-time time-averaged holographic interferometry, a hologram is first recorded on a photographic plate when the object is at rest. When the hologram after development and fixation is replaced exactly in the original recording position, the object then starts to vibrate. The reconstructed object wave and the object wave from the vibrating object will interfere with each other to form interference fringes in real time.

Assuming that, when the object is at rest, the complex amplitudes of the object and reference waves are denoted by $O = a_o \exp(\mathrm{i}\varphi_o)$ and $R = a_r \exp(\mathrm{i}\varphi_r)$ respectively, then when the object is at rest the intensity distribution recorded on a photographic plate can be expressed as

$$I = (O+R)(O+R)^* = (a_o^2 + a_r^2) + a_o a_r \exp[\mathrm{i}(\varphi_o - \varphi_r)] + a_o a_r \exp[-\mathrm{i}(\varphi_o - \varphi_r)] \tag{2.38}$$

Assuming that the exposure time is equal to T, that the amplitude transmittance is proportional to the exposure, and that the constant of proportionality is denoted by β, then the amplitude transmittance of the hologram after development and fixation can be written by

$$t = \beta I T = \beta T(a_o^2 + a_r^2) + \beta T a_o a_r \exp[\mathrm{i}(\varphi_o - \varphi_r)] + \beta T a_o a_r \exp[-\mathrm{i}(\varphi_o - \varphi_r)] \tag{2.39}$$

The above hologram is replaced exactly in the original recording position and illuminated simultaneously with the original object and reference waves. Assuming that the complex amplitude of object wave at any time is given by $O' = a_o \exp[i(\varphi_o + \delta)]$, then the complex amplitude of the object wave passing through the hologram can be expressed as

$$\begin{aligned} A = (O' + R)t = & \beta T(a_o^3 + a_o a_r^2)\exp[i(\varphi_o + \delta)] + \beta T a_o^2 a_r \exp[i(2\varphi_o - \varphi_r + \delta)] \\ & + \beta T a_o^2 a_r \exp[i(\varphi_r + \delta)] + \beta T(a_o^2 a_r + a_r^3)\exp(i\varphi_r) \\ & + \beta T a_o a_r^2 \exp(i\varphi_o) + \beta T a_o a_r^2 \exp[-i(\varphi_o - 2\varphi_r)] \end{aligned} \tag{2.40}$$

where the first term $\beta T(a_o^3 + a_o a_r^2)\exp[i(\varphi_o + \delta)]$ and the fifth term $\beta T a_o a_r^2 \exp(i\varphi_o)$ are related to the object wave, thus we only consider these two terms, i.e., we obtain

$$A' = \beta T \exp(i\varphi_o)[a_o a_r^2 + (a_o^3 + a_o a_r^2)\exp(i\delta)] \tag{2.41}$$

Using $I_o \ll I_r$, i.e. $a_o^2 \ll a_r^2$, Eq. (2.41) can be simplified as

$$A' = \beta T a_o a_r^2 \exp(i\varphi_o)[1 + \exp(i\delta)] \tag{2.42}$$

Therefore, the corresponding intensity distribution at any time is equal to

$$I' = A'A'^* = 2(\beta T a_o a_r^2)^2(1 + \cos\delta) \tag{2.43}$$

Substituting $\delta = k(\boldsymbol{e} - \boldsymbol{e}_0)\cdot \boldsymbol{A}\sin\omega t$ into Eq. (2.43), we have

$$I' = 2(\beta T a_o a_r^2)^2\{1 + \cos[k(\boldsymbol{e} - \boldsymbol{e}_0)\cdot \boldsymbol{A}\sin\omega t]\} \tag{2.44}$$

When the above intensity distribution is observed, the result will be a time-averaged value of the instantaneous intensity distribution. Assuming that the observation time is denoted by τ and that $\tau \gg 2\pi/\omega$, then the observed intensity distribution can be given by

$$I_\tau = \frac{1}{\tau}\int_0^\tau I'\mathrm{d}t = 2(\beta T a_o a_r^2)^2\left\{1 + \frac{1}{\tau}\int_0^\tau \cos[k(\boldsymbol{e} - \boldsymbol{e}_0)\cdot \boldsymbol{A}\sin\omega t]\mathrm{d}t\right\} = 2(\beta T a_o a_r^2)^2\{1 + J_0[k(\boldsymbol{e} - \boldsymbol{e}_0)\cdot \boldsymbol{A}]\} \tag{2.45}$$

Eq. (2.45) shows that the intensity distribution in real-time time-averaged holographic interferometry is related to $\{1 + J_0[k(\boldsymbol{e} - \boldsymbol{e}_0)\cdot \boldsymbol{A}]\}$. When $\{1 + J_0[k(\boldsymbol{e} - \boldsymbol{e}_0)\cdot \boldsymbol{A}]\}$ is maximum, bright fringes will be formed, and when $\{1 + J_0[k(\boldsymbol{e} - \boldsymbol{e}_0)\cdot \boldsymbol{A}]\}$ is minimum. minimum ,dark fringes will be formed Therefore, the amplitude distribution of vibration can be determined when the orders of fringes are determined. The $[1 + J_0(\alpha)]$-α curve is shown in Fig. 2.8.

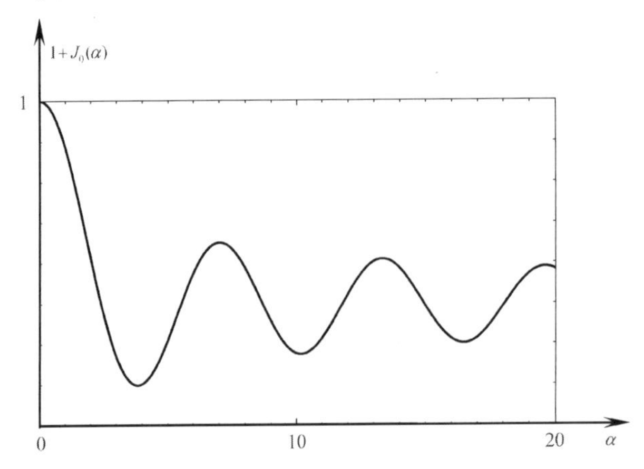

Fig. 2.8 $[1 + J_0(\alpha)]$-α curve

2.2.6 Stroboscopic Holographic Interferometry

In stroboscopic holographic interferometry, a vibrating object is illuminated sequentially by two laser pulses whose duration is much shorter than the period of vibration. Two series of exposure, respectively corresponding to two different states of the object, are recorded on a single photographic plate. Therefore, the hologram recorded in stroboscopic holographic interferometry is similar to a double-exposed hologram and will display interference fringes related to the phase change between these two vibration states.

Assuming that the complex amplitudes of two object waves at two different instantaneous states of the vibrating object are respectively denoted by $O_1 = a_o \exp[i(\varphi_o + \delta_1)]$ and $O_2 = a_o \exp[i(\varphi_o + \delta_2)]$, where a_o being the amplitude of the object wave, $(\varphi_o + \delta_1)$ and $(\varphi_o + \delta_2)$ the phases of the object wave at two different instantaneous states, and δ_1 and δ_2 the phase changes of the object wave caused due to the object deformation, and assuming that the complex amplitude of the reference wave is equal to $R = a_r \exp(i\varphi_r)$, then the intensity distribution recorded on a photographic plate corresponding to these two instantaneous states can given, respectively, by

$$\begin{aligned} I_1 &= (O_1+R)\cdot(O_1+R)^* = (a_o^2+a_r^2)+a_o a_r \exp[i(\varphi_o+\delta_1-\varphi_r)]+a_o a_r \exp[-i(\varphi_o+\delta_1-\varphi_r)] \\ I_2 &= (O_2+R)\cdot(O_2+R)^* = (a_o^2+a_r^2)+a_o a_r \exp[i(\varphi_o+\delta_2-\varphi_r)]+a_o a_r \exp[-i(\varphi_o+\delta_2-\varphi_r)] \end{aligned} \tag{2.46}$$

Assuming that the exposure time is τ_1 for the first instantaneous state and τ_2 for the second instantaneous state and that the numbers of exposure are denoted by N_1 for the first instantaneous state and N_2 for the second instantaneous state, the total exposure recorded on the photographic plate is equal to

$$\begin{aligned} E &= I_1 N_1 \tau_1 + I_2 N_2 \tau_2 \\ &= (a_o^2 + a_r^2)(N_1\tau_1 + N_2\tau_2) + a_o a_r \exp[i(\varphi_o - \varphi_r)][N_1\tau_1 \exp(i\delta_1) + N_2\tau_2 \exp(i\delta_2)] \\ &\quad + a_o a_r \exp[-i(\varphi_o - \varphi_r)][N_1\tau_1 \exp(-i\delta_1) + N_2\tau_2 \exp(-i\delta_2)] \end{aligned} \tag{2.47}$$

Assuming that the amplitude transmittance is proportional to the exposure, and that the constant of proportionality is denoted by β, then the amplitude transmittance of the hologram after development and fixation can be expressed as

$$\begin{aligned} t = \beta E &= \beta(a_o^2 + a_r^2)(N_1\tau_1 + N_2\tau_2) + \beta a_o a_r \exp[i(\varphi_o - \varphi_r)][N_1\tau_1 \exp(i\delta_1) + N_2\tau_2 \exp(i\delta_2)] \\ &\quad + \beta a_o a_r \exp[-i(\varphi_o - \varphi_r)][N_1\tau_1 \exp(-i\delta_1) + N_2\tau_2 \exp(-i\delta_2)] \end{aligned} \tag{2.48}$$

When the hologram is illuminated with the reference wave $R = a_r \exp(i\varphi_r)$, the complex amplitude of the light wave passing through the hologram can be described by

$$\begin{aligned} A = Rt &= \beta(a_o^2 a_r + a_r^3)(N_1\tau_1 + N_2\tau_2)\exp(i\varphi_r) \\ &\quad + \beta a_o a_r^2 \exp(i\varphi_o)[N_1\tau_1 \exp(i\delta_1) + N_2\tau_2 \exp(i\delta_2)] \\ &\quad + \beta a_o a_r^2 \exp[-i(\varphi_o - 2\varphi_r)][N_1\tau_1 \exp(-i\delta_1) + N_2\tau_2 \exp(-i\delta_2)] \end{aligned} \tag{2.49}$$

where the first term is the zeroth order diffraction wave, the second term the positive first order diffraction wave, and the third term the negative first order diffraction wave (i.e. conjugate wave).

If only the positive first order diffraction wave is considered, the complex amplitude passing through the hologram can be expressed as

$$A' = \beta a_o a_r^2 \exp(\mathrm{i}\varphi_o)[N_1\tau_1 \exp(\mathrm{i}\delta_1) + N_2\tau_2 \exp(\mathrm{i}\delta_2)] \tag{2.50}$$

The corresponding intensity distribution is equal to

$$I' = A'A'^* = (\beta a_o a_r^2)^2[(N_1\tau_1)^2 + (N_2\tau_2)^2 + 2N_1\tau_1 N_2\tau_2 \cos(\delta_2 - \delta_1)] \tag{2.51}$$

Assuming that the total exposure time for each instantaneous state is equal to T, i.e. $N_1\tau_1 = N_2\tau_2 = T$, then Eq. (2.51) can be simplified as

$$I' = 2(\beta T a_o a_r^2)^2[1 + \cos(\delta_2 - \delta_1)] \tag{2.52}$$

This is a general expression, and two particular cases of this expression will be discussed in detail.

(1) Assuming that two instantaneous states are respectively located at the equilibrium position and the amplitude position, i.e. $\delta_1 = 0$ and $\delta_2 = k(\boldsymbol{e} - \boldsymbol{e}_0)\cdot\boldsymbol{A}$, then Eq. (2.52) can be rewritten by

$$I' = 2(\beta T a_o a_r^2)^2\{1 + \cos[k(\boldsymbol{e} - \boldsymbol{e}_0)\cdot\boldsymbol{A}]\} \tag{2.53}$$

Eq. (2.53) shows, when the condition

$$(\boldsymbol{e} - \boldsymbol{e}_0)\cdot\boldsymbol{A} = n\lambda \quad (n = 0, \pm1, \pm2, \cdots) \tag{2.54}$$

is satisfied, bright fringes will be formed, and when

$$(\boldsymbol{e} - \boldsymbol{e}_0)\cdot\boldsymbol{A} = \left(n + \frac{1}{2}\right)\lambda \quad (n = 0, \pm1, \pm2, \cdots) \tag{2.55}$$

is met, dark fringes will be formed.

(2) Assuming that two instantaneous states are respectively located at the two out-of-phase amplitude positions, i.e. $\delta_1 = -k(\boldsymbol{e} - \boldsymbol{e}_0)\cdot\boldsymbol{A}$ and $\delta_2 = k(\boldsymbol{e} - \boldsymbol{e}_0)\cdot\boldsymbol{A}$, then Eq. (2.52) can be rewritten by

$$I' = 2(\beta T a_o a_r^2)^2\{1 + \cos[2k(\boldsymbol{e} - \boldsymbol{e}_0)\cdot\boldsymbol{A}]\} \tag{2.56}$$

Therefore, when

$$(\boldsymbol{e} - \boldsymbol{e}_0)\cdot\boldsymbol{A} = \frac{1}{2}n\lambda \quad (n = 0, \pm1, \pm2, \cdots) \tag{2.57}$$

is satisfied, bright fringes will be formed, and when

$$(\boldsymbol{e} - \boldsymbol{e}_0)\cdot\boldsymbol{A} = \frac{1}{2}\left(n + \frac{1}{2}\right)\lambda \quad (n = 0, \pm1, \pm2, \cdots) \tag{2.58}$$

is met, dark fringes will be formed.

Chapter 3 Speckle Photography and Speckle Interferometry

When a rough object with surface roughness on a scale of the wavelength of light is illuminated with a highly coherent light source (or laser), its surface will scatter numerous coherent wavelets. These scattered wavelets will interfere with each other in the space around the object to form numerous spots of brightness and darkness that are randomly distributed. These random spots with granular structure called speckles.

In fact, speckles have long been discovered, but have not been able to attract attention. Holography has been developed rapidly since the laser was invented, but the presence of speckles has greatly affected the quality of holography. Therefore it was not until then that speckles caused widespread interest. However, at that time speckles were studied as holographic noise and the purpose of research was to eliminate or reduce speckle noise in holography. With the intensive study on speckles, it was found that speckles could also be used for deformation measurement and vibration analysis.

When an object with optically rough surface, which is illuminated by laser light, is displaced or deformed, the distribution of speckles formed in the space around this object will also be moved or changed according to a certain law. Therefore, the displacement or deformation of the object being measured can be measured by analyzing the movement or change of speckles.

3.1 Speckle Photography

Speckle photography is formed by interference between random wavelets scattered from a rough object. Speckle photography usually includes double-exposure speckle photography, time-averaged speckle photography, stroboscopic speckle photography, etc.

3.1.1 Double-Exposure Speckle Photography

Double-exposure speckle photography requires two series of exposure of two instantaneous speckle fields; i.e., these two speckle fields are recorded on a single specklegram (i.e., speckle pattern). When this specklegram after development and fixation is subjected to filtering (i.e., pointwise filtering or whole field filtering), the displacement of the speckles recorded on the specklegram can be obtained, and the displacement or deformation of the object to be measured can then be calculated by using the displacement conversion relation between object and image points.

1. Specklegram Recording

The recording system used in double-exposure speckle photography is shown in Fig. 3.1. A beam of laser light (white light or partially coherent light can also be used in speckle photography) is used to illuminate the object plane, and the specklegram is recorded in the image plane. Two states of the object plane before and after deformation are recorded on a single specklegram through two series of exposure.

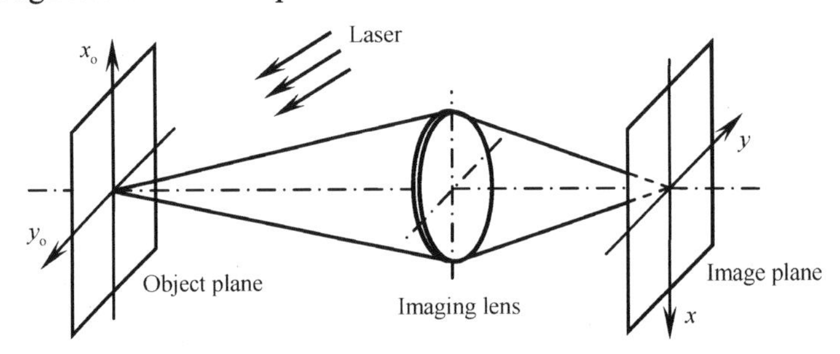

Fig. 3.1 Recording system

Assuming that the intensity distributions of the image plane before and after deformation are represented by $I_1(x,y)$ and $I_2(x,y)$, then we have

$$I_2(x,y) = I_1(x-u, y-v) \tag{3.1}$$

where $u = u(x,y)$ and $v = v(x,y)$ are the displacement components of point (x,y) on the specklegram along the x and y directions.

Assuming that the exposure time for each recording is equal to T, then the exposure of specklegram can be expressed as

$$E(x,y) = T[I_1(x,y) + I_2(x,y)] \tag{3.2}$$

Assuming that, within a certain range of exposure, the amplitude transmittance of the specklegram after development and fixation is proportional to its exposure, and that the constant of proportionality is denoted by β, then the amplitude transmittance of the specklegram can be given by

$$t(x,y) = \beta E(x,y) = \beta T[I_1(x,y) + I_2(x,y)] \tag{3.3}$$

2. Specklegram Filtering

When the specklegram is placed into a filtering system, as shown in Fig. 3.2, and illuminated by a collimated laser with unit amplitude, the spectral distribution on the Fourier transform plane can be written by

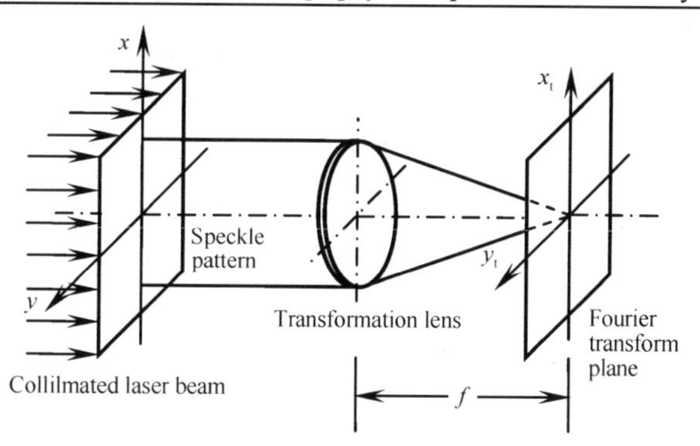

Fig. 3.2 Filtering system

$$\mathrm{FT}\{t(x,y)\} = \beta T[\mathrm{FT}\{I_1(x,y)\} + \mathrm{FT}\{I_2(x,y)\}] \tag{3.4}$$

where $\mathrm{FT}\{\cdot\}$ represents the Fourier transform.

By using the shift theorem of the Fourier transform (i.e., translation in the space domain introduces a linear phase shift in the frequency domain), we have

$$\mathrm{FT}\{I_2(x,y)\} = \mathrm{FT}\{I_1(x-u,y-v)\} = \mathrm{FT}\{I_1(x,y)\}\exp[-\mathrm{i}2\pi(uf_x + vf_y)] \tag{3.5}$$

where $(f_x, f_y) = (x_t, y_t)/(\lambda f)$ are respectively the frequency coordinates along the x and y directions on the Fourier transform plane, with λ being the wavelength of laser, f the focal length of the transformation lens, and (x_t, y_t) the coordinates along the x and y directions on the Fourier transform plane.

Using Eq. (3.5), then Eq. (3.4) can be rewritten by

$$\mathrm{FT}\{t(x,y)\} = \beta T \mathrm{FT}\{I_1(x,y)\}\{1 + \exp[-\mathrm{i}2\pi(uf_x + vf_y)]\} \tag{3.6}$$

Therefore, after the specklegram is subjected to the Fourier transform, the intensity distribution of diffraction halo in the Fourier transform plane is equal to

$$I(f_x, f_y) = |\mathrm{FT}\{t(x,y)\}|^2 = 2\beta^2 T^2 |\mathrm{FT}\{I_1(x,y)\}|^2 \{1 + \cos[2\pi(uf_x + vf_y)]\} \tag{3.7}$$

It can be seen from Eq. (3.7) that cosine interference fringes will appear in the diffraction halo.

If displacement vectors of various points on the specklegram are identical, then interference fringes will appear on the Fourier transform plane. However, the displacement vectors of various points on the specklegram are usually different from each other. At this time, the interference fringes with different spacings and directions will be superposed on the Fourier transform plane, therefore these interference fringes can not be observed directly, but can be revealed by pointwise or wholefield filtering.

3. Pointwise Filtering

Consider that point P on the specklegram is illuminated with a thin laser beam, as shown in Fig. 3.3. Assuming that the illuminated area is small, then interference fringes, i.e., Young fringes, can be displayed directly on the observation plane.

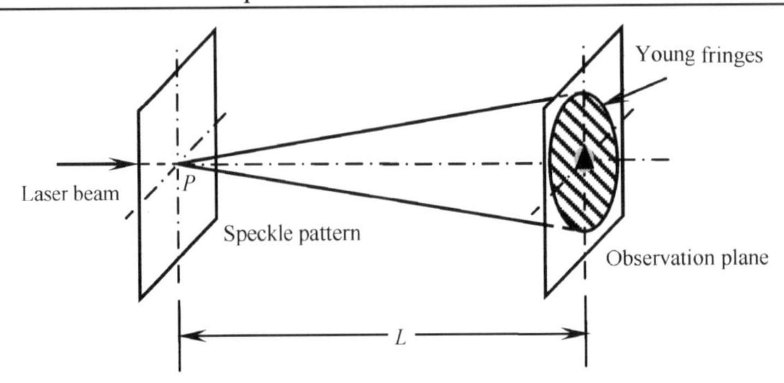

Fig. 3.3 Pointwise filtering

Since the speckle displacement is always perpendicular to the Young fringes in direction and inversely proportional to the spacing of two adjacent Young fringes in magnitude, the magnitude of the speckle displacement at the point P on the specklegram can be expressed as

$$d = \frac{\lambda L}{\varDelta} \tag{3.8}$$

where $\varDelta$ is the spacing of adjacent Young fringes, and L is the distance of the observation plane from the specklegram.

4. Wholefield Filtering

The wholefield filtering system is shown in Fig. 3.4. When an opaque screen with a filtering hole is placed into the Fourier transform plane, interference fringes can be observed through the filtering hole. The spacing of interference fringes is changed continuously when the filtering hole moves along the radial direction and is decreased continuously when the filtering hole is far away from the optical axis. The direction of interference fringes is also changed continuously when the filtering hole moves along the circumferential direction.

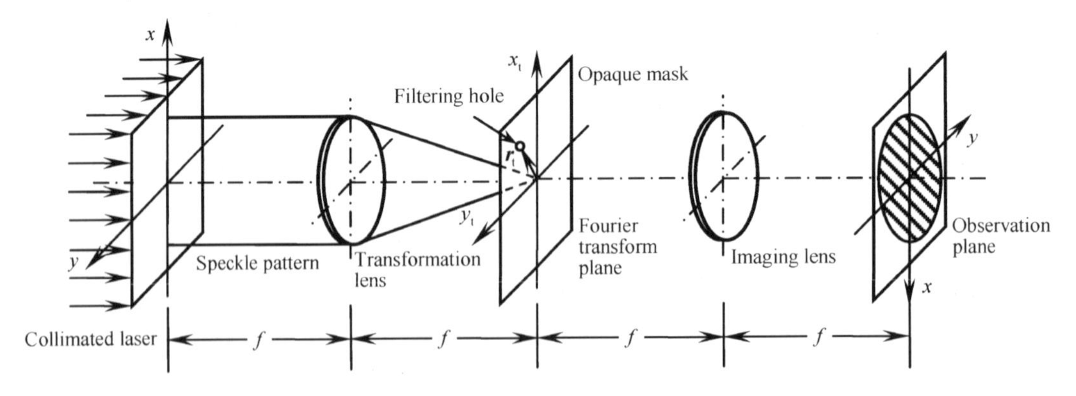

Fig. 3.4 Wholefield filtering

The interference fringes obtained in wholefield filtering represent the displacement contours of various points on the specklegram along the filtering hole. When the filtering hole is located at $(x_t, 0)$, from Eq. (3.7) the condition for producing bright fringes is

$$u = \frac{m}{f_x} = \frac{m\lambda f}{x_t} \quad (m = 0, \pm 1, \pm 2, \cdots) \tag{3.9}$$

Similarly, when the filtering hole is located at $(0, y_t)$, and the condition

$$u = \frac{m}{f_x} = \frac{m\lambda f}{x_t} \quad (m = 0, \pm 1, \pm 2, \cdots) \tag{3.10}$$

is satisfied, bright fringes will also be generated.

3.1.2 Time-Averaged Speckle Photography

Time-averaged speckle photography is to record continuously the specklegram of a oscillating object subjected to steady state vibration with a long exposure time much greater than the period of vibration. Therefore, this method can record all the vibration states of the oscillating object into a single specklegram, and the interference fringes characterizing the amplitude distribution of the oscillating object can be obtained by filtering the specklegram after development and fixation.

1. Specklegram Recording

Assume that the intensity distribution on the image plane at any time is denoted by $I(x-u, y-v)$, where $u = u(x, y; t)$ and $v = v(x, y; t)$ are the displacement components at the time t of point (x, y) on the specklegram along the x and y directions, respectively. Assuming that the exposure time is equal to T, the exposure of the specklegram can be expressed as

$$E(x, y) = \int_0^T I(x-u, y-v)\mathrm{d}t \tag{3.11}$$

Assuming that, within a certain range of exposure, the amplitude transmittance of the specklegram after development and fixation is proportional to its exposure, and that the constant of proportionality is denoted by β, then the amplitude transmittance of the specklegram can be given by

$$t(x, y) = \beta E(x, y) = \beta \int_0^T I(x-u, y-v)\mathrm{d}t \tag{3.12}$$

2. Specklegram Filtering

When the specklegram placed into a wholefield filtering system is illuminated by a beam of collimated laser with unit amplitude, then the spectral distribution on the Fourier transform plane can be expressed as

$$\mathrm{FT}\{t(x, y)\} = \beta \int_0^T \mathrm{FT}\{I(x-u, y-v)\}\mathrm{d}t \tag{3.13}$$

From the shift theorem of the Fourier transform, we have

$$\mathrm{FT}\{I(x-u,y-v)\}=\mathrm{FT}\{I(x,y)\}\exp[-\mathrm{i}2\pi(uf_x+vf_y)] \tag{3.14}$$

Using Eq. (3.14), then Eq. (3.13) can be rewritten as

$$\mathrm{FT}\{t(x,y)\}=\beta\mathrm{FT}\{I(x,y)\}\int_0^T \exp[-\mathrm{i}2\pi(uf_x+vf_y)]\mathrm{d}t \tag{3.15}$$

Therefore, when the specklegram is subjected to the Fourier transform, the intensity distribution of diffraction halo on the Fourier transform plane is equal to

$$I(f_x,f_y)=|\mathrm{FT}\{t(x,y)\}|^2=\beta^2\,|\mathrm{FT}\{I(x,y)\}|^2\left|\int_0^T \exp[-\mathrm{i}2\pi(uf_x+vf_y)]\mathrm{d}t\right|^2 \tag{3.16}$$

Assuming that the object is in harmonic vibration, the x and y displacement components at time t of any point on the specklegram can be expressed as

$$u=A_x\sin\omega t,\quad v=A_y\sin\omega t \tag{3.17}$$

where $A_x=A_x(x,y)$ and $A_y=A_y(x,y)$ are the amplitude components of point (x,y) on the specklegram respectively along the x and y directions, and ω is the circular or angular frequency of the vibrating object. Substituting Eq. (3.17) into Eq. (3.16), we obtain

$$I(f_x,f_y)=\beta^2\,|\mathrm{FT}\{I(x,y)\}|^2\left|\int_0^T \exp[-\mathrm{i}2\pi(A_xf_x+A_yf_y)\sin\omega t]\mathrm{d}t\right|^2 \tag{3.18}$$

Assuming that $T\gg 2\pi/\omega$, Eq. (3.18) can be rewritten by

$$I(f_x,f_y)=\beta^2T^2\,|\mathrm{FT}\{I(x,y)\}|^2\,J_0^2[2\pi(A_xf_x+A_yf_y)] \tag{3.19}$$

where J_0 is the zero order Bessel function of the first kind. It can be seen from the above equation that the interference fringes related to the amplitude distribution will appear in the diffraction halo. The amplitude vectors of various points on the specklegram are usually different from each other, i.e., the interference fringes having different spacings and directions will be superposed on the Fourier transform plane, therefore these interference fringes can not be observed directly on the Fourier transform plane. However, they can be extracted by wholefield filtering.

It can be seen from Eq. (3.19) that bright fringes will be formed when $J_0^2[2\pi(A_xf_x+A_yf_y)]$ takes maximum values. In particular, when the maximum value of $J_0^2[2\pi(A_xf_x+A_yf_y)]$ is taken at $A_x=A_y=0$, the most bright fringe (i.e., nodal line of vibration) can be obtained. Similarly, dark fringes will be formed when $J_0^2[2\pi(A_xf_x+A_yf_y)]$ takes minimum values, or when the condition

$$2\pi(A_xf_x+A_yf_y)=\alpha\quad(\alpha=2.41,\ 5.52,\ 8.65,\ 11.79,\ 14.98,\ \cdots) \tag{3.20}$$

is satisfied. The $J_0^2(\alpha)$-α curve is shown in Fig. 3.5.

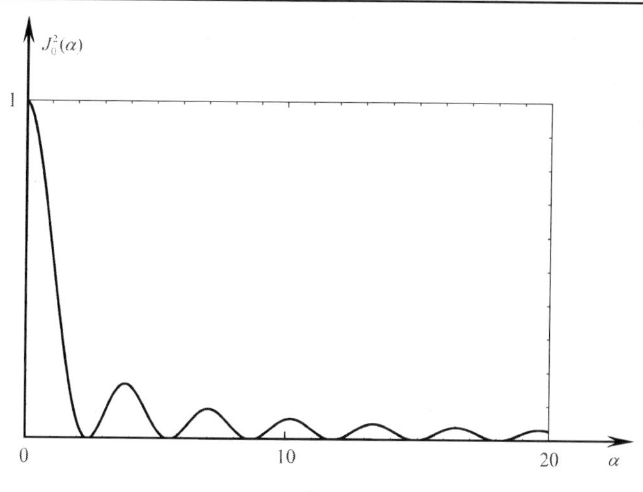

Fig. 3.5 $J_0^2(\alpha)$ - α curve

When the filtering hole is located at $(x_t, 0)$ or $(0, y_t)$, from Eq. (3.19) the conditions for appearing dark fringes can be given, respectively, by

$$\begin{aligned} A_x &= \frac{\alpha_x}{2\pi f_x} = \frac{\alpha_x \lambda f}{2\pi x_t} \quad (\alpha_x = 2.41,\ 5.52,\ 8.65,\ 11.79,\ 14.98,\ \cdots) \\ A_y &= \frac{\alpha_y}{2\pi f_y} = \frac{\alpha_y \lambda f}{2\pi y_t} \quad (\alpha_y = 2.41,\ 5.52,\ 8.65,\ 11.79,\ 14.98,\ \cdots) \end{aligned} \tag{3.21}$$

3.1.3 Stroboscopic Speckle Photography

Two instantaneous states of the object undergoing dynamic deformation are recorded onto a single specklegram in stroboscopic speckle photography. The interference fringes corresponding to these two instantaneous states can be obtained by filtering the specklegram after development and fixation, and a dynamic measurement can be carried out by analyzing these interference fringes.

1. Specklegram Recording

Assuming that the intensity distributions of the image plane, corresponding to two instantaneous states of the object undergoing dynamic deformation, are respectively denoted by $I(x-u_1, y-v_1)$ and $I(x-u_2, y-v_2)$, and that the exposure time for these two instantaneous states is denoted by τ, then the exposure of the specklegram can be expressed as

$$E(x,y) = \tau[I(x-u_1, y-v_1) + I(x-u_2, y-v_2)] \tag{3.22}$$

where $u_1 = u_1(x,y;t_1)$ and $v_1 = v_1(x,y;t_1)$ are the displacement components at time t_1 along the x and y directions of point (x,y) on the specklegram, and $u_2 = u_2(x,y;t_2)$ and $v_2 = v_2(x,y;t_2)$ are the displacement components at time t_2 along the x and y directions of point (x,y) on the same specklegram.

Assuming that, within a certain range of exposure, the amplitude transmittance of the specklegram after development and fixation is proportional to its exposure, and that the constant of proportionality is denoted by β, then the amplitude transmittance of the specklegram can be given by

$$t(x,y)=\beta E(x,y)=\beta\tau[I(x-u_1,y-v_1)+I(x-u_2,y-v_2)] \tag{3.23}$$

2. Specklegram Filtering

When the specklegram is subjected to wholefield filtering and illuminated by a beam of collimated laser with unit amplitude, then the spectral distribution on the Fourier transform plane can be expressed as

$$\mathrm{FT}\{t(x,y)\}=\beta\tau[\mathrm{FT}\{I(x-u_1,y-v_1)\}+\mathrm{FT}\{I(x-u_2,y-v_2)\}] \tag{3.24}$$

Using the shift theorem of the Fourier transform, we obtain

$$\begin{aligned}\mathrm{FT}\{I(x-u_1,y-v_1)\}&=\mathrm{FT}\{I(x,y)\}\exp[-\mathrm{i}2\pi(u_1f_x+v_1f_y)]\\ \mathrm{FT}\{I(x-u_2,y-v_2)\}&=\mathrm{FT}\{I(x,y)\}\exp[-\mathrm{i}2\pi(u_2f_x+v_2f_y)]\end{aligned} \tag{3.25}$$

Using Eq. (3.25), Eq. (3.24) can be expressed as

$$\mathrm{FT}\{t(x,y)\}=\beta\tau\mathrm{FT}\{I(x,y)\}\{\exp[-\mathrm{i}2\pi(u_1f_x+v_1f_y)]+\exp[-\mathrm{i}2\pi(u_2f_x+v_2f_y)]\} \tag{3.26}$$

Therefore, when the specklegram is subjected to the Fourier transform, the intensity distribution of the diffraction halo on the Fourier transform plane can be given by

$$I(f_x,f_y)=|\mathrm{FT}\{t(x,y)\}|^2=2\beta^2\tau^2\,|\mathrm{FT}\{I(x,y)\}|^2\,\{1+\cos[2\pi(uf_x+vf_y)]\} \tag{3.27}$$

where $u=u_2(x,y;t_2)-u_1(x,y;t_1)$ and $v=v_2(x,y;t_2)-v_1(x,y;t_1)$ are the relative displacement components of point (x,y) on the specklegram at two instantaneous times t_1 and t_2 along the x and y directions, respectively. It can be seen from the above equation that interference fringes can be extracted by pointwise or wholefield filtering. When the wholefield filtering is adopted and the filtering hole is located at $(x_t,0)$ or $(0,y_t)$, the conditions for producing bright fringes are respectively given by

$$u=\frac{m}{f_x}=\frac{m\lambda f}{x_t}\quad(m=0,\ \pm1,\ \pm2,\ \cdots),\qquad v=\frac{n}{f_y}=\frac{n\lambda f}{y_t}\quad(n=0,\ \pm1,\ \pm2,\ \cdots) \tag{3.28}$$

3.2 Speckle Interferometry

Speckle interferometry is formed by interference between a speckle field scattered from the object with rough surface and a reference light.

3.2.1 In-Plane Displacement Measurement

The in-plane displacement measurement system is shown in Fig. 3.6. Two beams of laser located in the plane perpendicular to the y_o axis are used to illuminate symmetrically the object plane. Two series of exposure, one before and the other after deformation, are recorded on a photographic plate. The recorded specklegram will generate contour fringes characterizing the in-plane displacement component along the x direction when it is subjected to filtering.

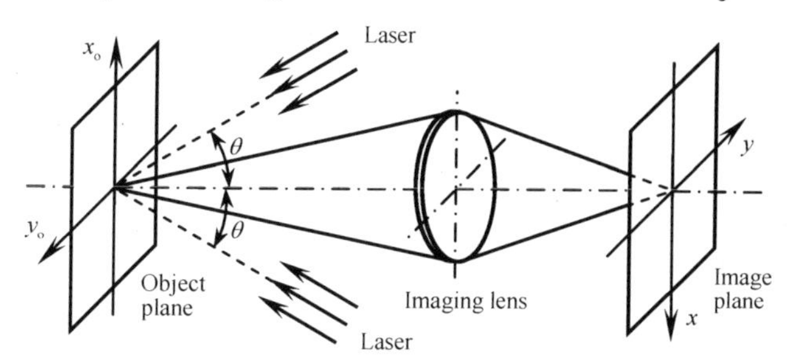

Fig. 3.6 In-plane displacement measurement system

The intensity distribution on the image plane before deformation can be expressed as

$$I_1(x,y) = I_{o1} + I_{o2} + 2\sqrt{I_{o1}I_{o2}}\cos\varphi \tag{3.29}$$

where I_{o1} and I_{o2} are the intensity distributions of two incident light waves, and φ is the phase difference between these two incident light waves.

Similarly, the intensity distribution on the image plane after deformation can be expressed as

$$I_2(x,y) = I_{o1} + I_{o2} + 2\sqrt{I_{o1}I_{o2}}\cos(\varphi + \delta) \tag{3.30}$$

where $\delta = \delta_1 - \delta_2$ with δ_1 and δ_2 being the phase changes of two incident light waves caused by the deformation of object. According to Eq. (2.8), δ_1 and δ_2 can be expressed, respectively, as

$$\delta_1 = k[w_o(1+\cos\theta) + u_o\sin\theta], \quad \delta_2 = k[w_o(1+\cos\theta) - u_o\sin\theta] \tag{3.31}$$

where k is the wave number of the laser used, θ is the angle between each of the incident lights and the optical axis, and u_o and w_o are the displacement components on the object plane along the x and z directions, respectively. Thus, the relative phase change of two incident light waves caused by the deformation of object is

$$\delta = 2ku_o\sin\theta \tag{3.32}$$

Assuming that the exposure time before and after deformation is denoted by T, and that a linear relation exists between the amplitude transmittance and the exposure, then the amplitude transmittance of specklegram can be given by

$$\begin{aligned} t(x,y) &= \beta T[I_1(x,y) + I_2(x,y)] \\ &= 2\beta T(I_{o1} + I_{o2}) + 4\beta T\sqrt{I_{o1}I_{o2}}\cos\left(\varphi + \frac{1}{2}\delta\right)\cos\left(\frac{1}{2}\delta\right) \end{aligned} \tag{3.33}$$

where β is a constant of proportionality, φ is the random function varying rapidly, and δ is the function varying slowly. The cosine function involving both φ and δ is a high frequency component, whereas the cosine function involving δ alone is a low frequency component. Thus, when the condition

$$\cos\left(\frac{1}{2}\delta\right)=0 \tag{3.34}$$

i.e.,

$$\delta=(2n+1)\pi \quad (n=0,\pm1,\pm2,\cdots) \tag{3.35}$$

is satisfied, dark fringes will be produced. Using Eq. (3.32), we have

$$u_o=\frac{(2n+1)\pi}{2k\sin\theta}=\frac{(2n+1)\lambda}{4\sin\theta} \quad (n=0,\pm1,\pm2,\cdots) \tag{3.36}$$

Due to the existence of the background intensity $2\beta T(I_{o1}+I_{o2})$, interference fringes can not be seen directly from the specklegram. When the specklegram is subjected to wholefield filtering, the contour fringes of in-plane displacement component can be revealed on the observation plane by observing the diffraction halo when the low frequency component has been filtered out.

3.2.2 Out-of-Plane Displacement Measurement

The out-of-plane displacement measurement system is shown in Fig. 3.7. The intensity distribution on the image plane is formed due to interference between the reference wave and the speckle field scattered from the object plane. Two series of exposure are respectively performed before and after deformation, and the specklegram obtained will produce contour fringes of out-of-plane displacement when it is subjected to filtering.

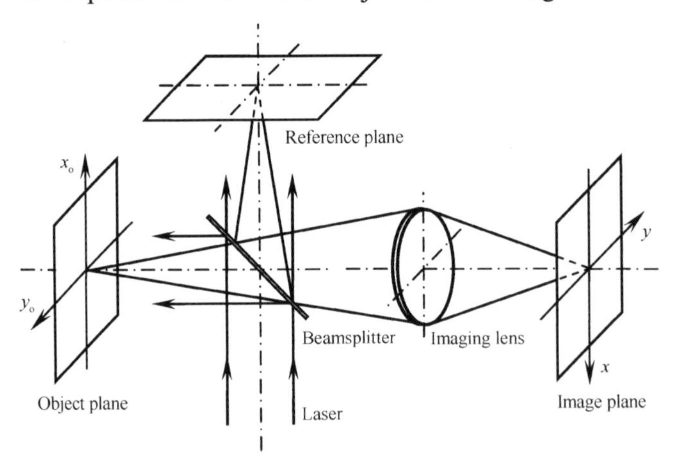

Fig. 3.7 Out-of-plane displacement measurement system

The intensity distribution corresponding to the first exposure is

$$I_1(x,y)=I_o+I_r+2\sqrt{I_oI_r}\cos\varphi \tag{3.37}$$

where I_o and I_r are respectively the intensity distributions of the object and reference waves, and φ is the phase difference between the object and reference waves.

Similarly, the intensity distribution corresponding to the second exposure is

$$I_2(x,y)=I_o+I_r+2\sqrt{I_oI_r}\cos(\varphi+\delta) \tag{3.38}$$

where δ is the relative phase change of the object and reference waves produced due to deformation. When the illumination and reception directions of the object wave are both perpendicular to the object plane, the phase change δ can be expressed as

$$\delta=2kw_o \tag{3.39}$$

where w_o is the out-of-plane displacement.

The intensity distribution of the specklegram subjected to two series of exposure can be expressed as

$$I(x,y)=I_1(x,y)+I_2(x,y)=2(I_o+I_r)+4\sqrt{I_oI_r}\cos\left(\varphi+\frac{1}{2}\delta\right)\cos\left(\frac{1}{2}\delta\right) \tag{3.40}$$

where $\cos\left(\varphi+\frac{1}{2}\delta\right)$ is a high frequency component, and $\cos\left(\frac{1}{2}\delta\right)$ is a low frequency component. When the condition

$$\delta=(2n+1)\pi \quad (n=0,\pm1,\pm2,\cdots) \tag{3.41}$$

is satisfied, dark fringes will be formed. Using Eq. (3.39), we obtain

$$w_o=\frac{(2n+1)\pi}{2k}=\frac{(2n+1)\lambda}{4} \quad (n=0,\pm1,\pm2,\cdots) \tag{3.42}$$

Due to the existence of the background intensity, interference fringes can not be seen directly from the specklegram. After filtering out the low frequency component, the contour fringes corresponding to the out-of-plane displacement component can be observed.

3.3 Speckle Shearing Interferometry

Speckle shearing interferometry is formed by interference between two mutually sheared speckle fields scattered from the same rough object surface. Speckle shearing interferometry can be used for measuring out-of-plane displacement derivatives (i.e., slopes).

The out-of-plane displacement derivative measurement system is shown in Fig. 3.8. A double-hole screen is placed in front of the imaging lens with two holes being along the x axis, and a shearing wedge is placed onto one of the two holes with the shearing direction along the x axis. When the object plane is illuminated with a beam of laser, then a point on the object plane will be imaged to two points on the image plane; or, two adjacent points on the object plane will be imaged to a single point on the image plane. The first exposure is performed before deformation, and the second after deformation. When the specklegram is subjected to filtering, the contour fringes of the out-of-plane displacement derivative along the double-hole direction will be formed.

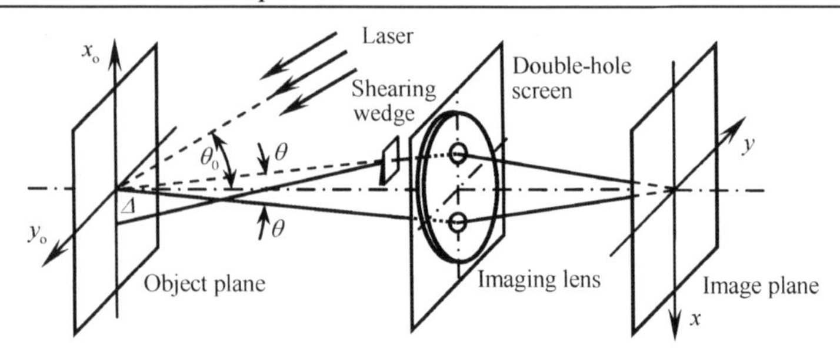

Fig. 3.8　Speckle shearing interferometric system

The shearing value on the object plane produced by the shearing wedge placed in front of the imaging lens can be expressed as

$$\varDelta = d_o(\mu - 1)\alpha \tag{3.43}$$

where d_o is the distance of the shearing wedge from the object plane, and μ and α the refractive index and wedge angle of the shearing wedge.

Assuming that the intensity distributions of the light waves through two holes are I_{o1} and I_{o2}, then the intensity distribution of the image plane before deformation is

$$I_1(x, y) = I_{o1} + I_{o2} + 2\sqrt{I_{o1}I_{o2}}\cos(\varphi + \beta) \tag{3.44}$$

where φ is the relative phase corresponding to two object points, β is the phase of a grating structure produced due to interference of light waves through two holes.

Similarly, the intensity distribution of the image plane after deformation is

$$I_2(x, y) = I_{o1} + I_{o2} + 2\sqrt{I_{o1}I_{o2}}\cos(\varphi + \delta + \beta) \tag{3.45}$$

where $\delta = \delta_1 - \delta_2$ with δ_1 and δ_2 being the phase change of the light waves through two holes caused by the deformation of object. According to the directions of illumination and observation, δ can be expressed as

$$\delta = k\left\{2u_o\sin\theta + \left[(\sin\theta + \sin\theta_0)\frac{\partial u_o}{\partial x} + (\cos\theta + \cos\theta_0)\frac{\partial w_o}{\partial x}\right]\varDelta\right\} \tag{3.46}$$

where θ_0 and θ are the included angle between the illumination direction and the optical axis and the included angle between the observation direction and the optical axis, u_o is the in-plane displacement component along the x direction, and $\partial u_o/\partial x$ and $\partial w_o/\partial x$ are the in-plane displacement derivative and the out-of-plane displacement derivative, both with respect to x. Using $\sin\theta << 1$, Eq. (3.46) can be simplified as

$$\delta = k\left[\sin\theta_0\frac{\partial u_o}{\partial x} + (1 + \cos\theta_0)\frac{\partial w_o}{\partial x}\right]\varDelta \tag{3.47}$$

When the laser beam is perpendicular to the object plane, i.e., $\theta_0 = 0$, Eq. (3.47) can be rewritten by

$$\delta = 2k\frac{\partial w_{\mathrm{o}}}{\partial x}\Delta \tag{3.48}$$

After two series of exposure, the intensity distribution on the image plane can be given by

$$I(x,y) = I_1(x,y) + I_2(x,y) = 2(I_{\mathrm{o}1} + I_{\mathrm{o}2}) + 2\sqrt{I_{\mathrm{o}1}I_{\mathrm{o}2}}[\cos(\varphi+\beta) + \cos(\varphi+\delta+\beta)] \tag{3.49}$$

When the specklegram is subjected to filtering and the first-order diffraction halo is allowed to pass through the filtering hole, the intensity distribution on the observation plane can be expressed as

$$I(x_{\mathrm{t}},y_{\mathrm{t}}) = 2I_{\mathrm{o}1}I_{\mathrm{o}2}(1+\cos\delta) \tag{3.50}$$

Thus, when the condition $\delta = 2n\pi$, i.e.,

$$\frac{\partial w_{\mathrm{o}}}{\partial x} = \frac{n\pi}{k\Delta} = \frac{n\lambda}{2\Delta} \quad (n = 0, \pm 1, \pm 2, \cdots) \tag{3.51}$$

is met, bright fringes can be formed on the observation plane.

Chapter 4 Geometric Moiré and Moiré Interferometry

A moiré pattern refers to a wavy pattern related to silk fabrics made in China. The ancient Chinese extruded two layers of silk to form moiré patterns on silk fabrics, and European silk dealers imported these silk fabrics and started to use the term "moiré".

4.1 Geometric Moiré

When two periodic structures, such as screens, gratings, etc., are superposed, a moiré pattern consisting of alternating fringes of brightness and darkness will be formed within the overlap region of the structures. These fringes of brightness and darkness appearing in this moiré pattern are called moiré fringes.

Because moiré fringes are extremely sensitive to a slight deformation or rotation of two overlapped structures, the moiré method has found a vast number of applications in many different fields. For example, it can be used for measuring in-plane and out-of-plane displacements based on moiré fringes formed when two gratings are superposed.

The fundamental measuring element used in geometric moiré is an amplitude grating consisting of opaque and transparent lines, as shown in Fig. 4.1. These opaque lines are called grating lines. The vertical distance between two adjacent grating lines is called the pitch of grating.

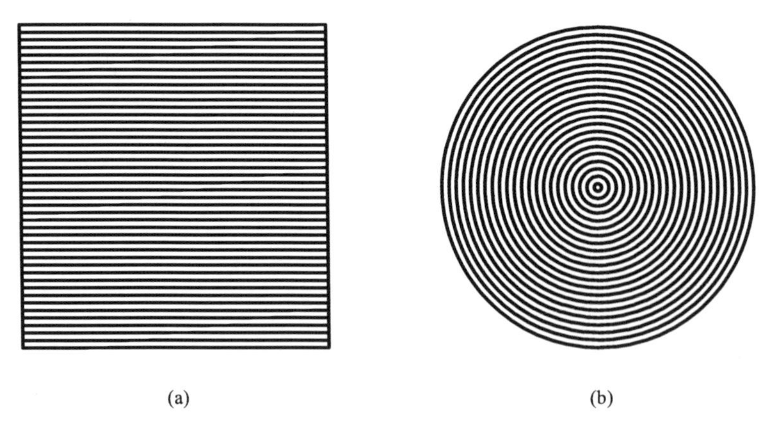

Fig. 4.1 Gratings

In geometric moiré, a commonly-used grating usually has rectilinear grating lines, Fig. 4.1(a), but grating lines can also be curvilinear, Fig. 4.1(b).

4.1.1 Geometric Moiré Formation

When two identical gratings are superimposed, the overlap region looks like a single grating and no moiré fringe appears in the overlap region if the grating lines of two superposed gratings completely coincide with each other, Fig. 4.2(a). However, when two superposed gratings have a relative rotation, Fig. 4.2(b), or when any of two superposed gratings is in tension, Fig. 4.2(c) or in compression, Fig. 4.2(d), a moiré pattern consisting of alternating fringes of brightness and darkness will be formed within the overlap region, although it does not appear in any of the original gratings.

When the grating lines of two superposed gratings have a continuous relative rotation or the grating pitches of two superposed gratings have a continuous relative change, the moiré fringes formed within the overlap region. will change continuously. Therefore, the displacement distribution can be obtained by observing these moiré fringes.

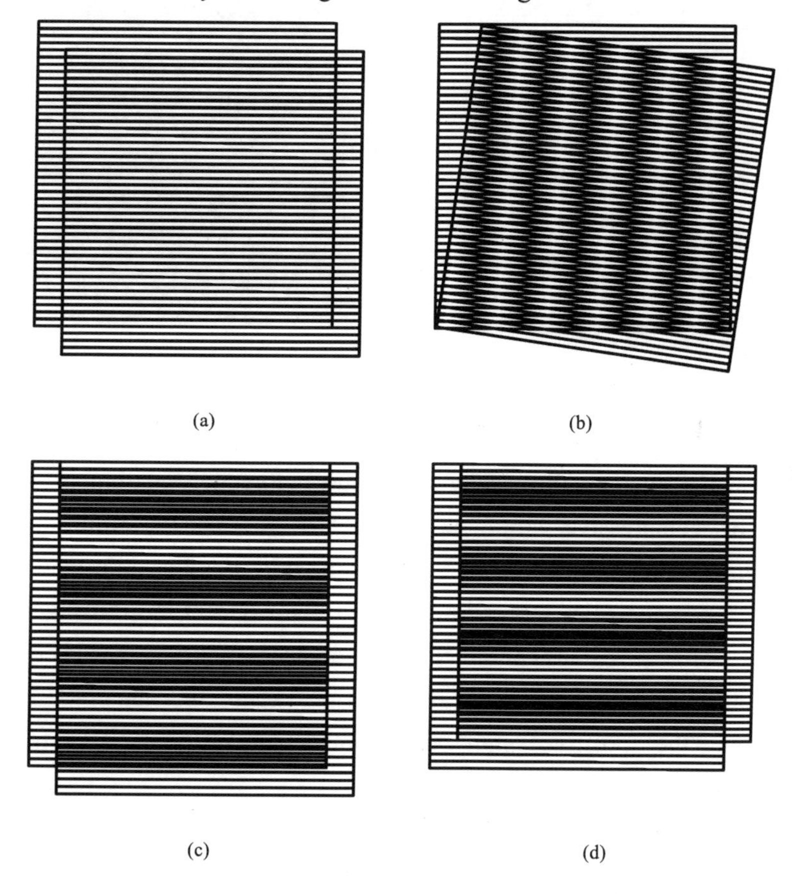

Fig. 4.2 Formation of moiré fringes

The moiré method usually requires two gratings; one, called the specimen grating, is affixed to the region being measured and will deform with the object, whereas the other, called the reference grating, coincides with the specimen grating but will not deform with the object.

4.1.2 Geometric Moiré for Strain Measurement

1. Tensile or Compressive Strain Measurement

1) Parallel Moiré Method

When the grating lines of the specimen and reference gratings having the same pitch are perpendicular to the tensile or compressive direction of the object to be measured, no moiré fringe will appear in the overlap region if the grating lines of the reference grating coincide with those of the specimen grating. However, when the object is subjected to a uniformly tensile or compressive deformation, the specimen grating will have the same deformation as the object. Therefore, moiré fringes will be formed within the overlap region of two superposed gratings, as shown in Fig. 4.3.

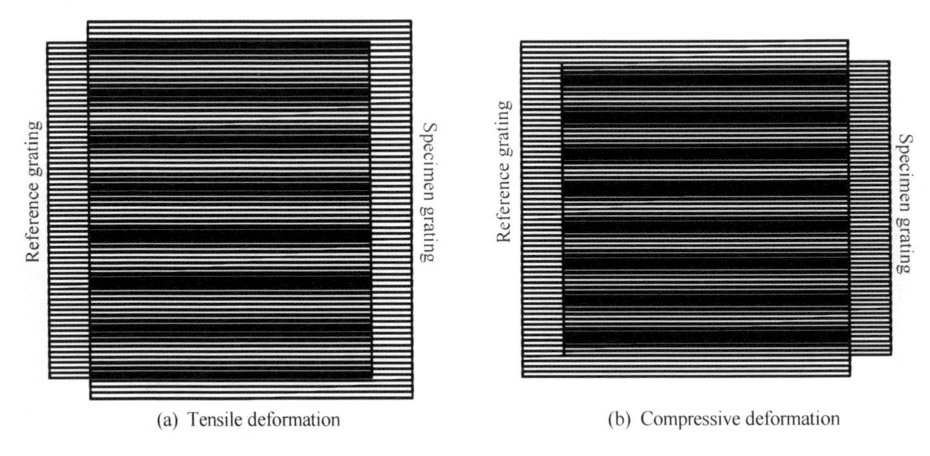

(a) Tensile deformation (b) Compressive deformation

Fig. 4.3 Parallel moiré

Assuming that the pitch of the specimen grating before deformation is p and that the uniform tensile or compressive strain is ε ($\varepsilon>0$ for tensile strain, and $\varepsilon<0$ for compressive strain), then the pitch of the specimen grating after deformation is equal to $p'=(1+\varepsilon)p$. If the perpendicular separation between two adjacent moiré fringes is denoted by f , then the number of grating lines contained in two adjacent moiré fringes is equal to $n=f/p$ before deformation and $n'=f/p\mp1$ after deformation, where the negative sign is used for tensile strain and the positive sign for compressive strain. Therefore, we have $f=n'p'=(f/p\mp1)(1+\varepsilon)p$, i.e.,

$$\varepsilon=\frac{\pm p}{f\mp p} \tag{4.1}$$

Using $p\ll f$, we can write

$$\varepsilon=\pm\frac{p}{f} \tag{4.2}$$

where the positive sign corresponds to tensile strain and the negative sign corresponds to compressive strain. Therefore, when the pitch p is known, the uniformly tensile or compressive strain perpendicular to the grating lines can be determined by using the perpendicular separation f of two adjacent moiré fringes.

2) Angular Moiré Method

Assume that grating lines of the specimen grating are perpendicular to the tensile or compressive direction of the object and that the reference grating has an included angle θ with respect to the specimen grating (θ is positive anticlockwise), as shown in Fig. 4.4. If the pitches are respectively q for the reference grating, p for the specimen grating before deformation, and p' for the specimen grating after deformation, then the strain can be expressed as

$$\varepsilon = \frac{p'-p}{p} = \frac{p'}{p} - 1 = \frac{q}{p}\frac{p'}{q} - 1 \tag{4.3}$$

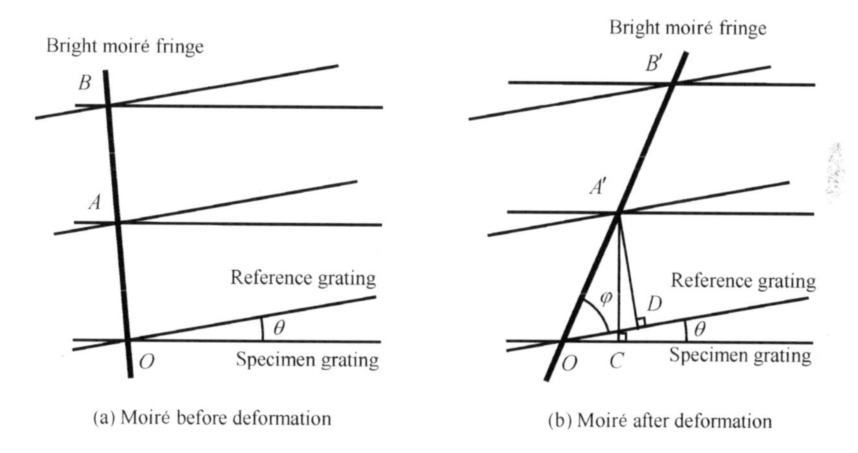

Fig. 4.4 Angular moiré

Assume that OAB denotes a bright moiré fringe before deformation and this bright moiré fringe after deformation is represented by $OA'B'$. If this bright moiré fringe after deformation has an included angle φ relative to the grating lines of the reference grating (φ is positive for counterclockwise rotation) as shown in Fig. 4.4, we then have

$$p' = A'C = OA'\sin(\theta+\varphi), \quad q = A'D = OA'\sin\varphi \tag{4.4}$$

Substituting Eq. (4.4) into Eq. (4.3), we obtain

$$\varepsilon = \frac{p'-p}{p} = \frac{p'}{p} - 1 = \frac{q}{p}\frac{\sin(\theta+\varphi)}{\sin\varphi} - 1 \tag{4.5}$$

Therefore, when p, q and θ are known, ε can be determined by measuring φ.

If the reference and specimen gratings have the same pitch, then Eq. (4.5) can be simplified as

$$\varepsilon = \frac{\sin(\theta+\varphi)}{\sin\varphi} - 1 \tag{4.6}$$

2. Shearing Strain Measurement

Shearing strain measurement is usually divided into two steps. In the first step, the grating lines of the reference and specimen gratings with the same pitch p are parallel to the x axis, as shown in Fig. 4.5. When the specimen grating is subjected to a shearing deformation, the pitch p of the specimen grating remains unchanged, the angle θ_y only makes the grating lines of the specimen grating move along the x direction, and the angle θ_x can make the grating lines of the specimen grating rotate. Therefore, only θ_x can cause moiré fringes. Using $BC \perp AC$, as shown in Fig. 4.5, we have

$$\sin\theta_x = \frac{BC}{AB} = \frac{p}{f_x} \tag{4.7}$$

where f_x is the spacing of two adjacent moiré fringes in the x direction (i.e. the direction of grating lines). Using $\sin\theta_x \approx \theta_x$, Eq. (4.7) can be rewritten by

$$\theta_x = \frac{p}{f_x} \tag{4.8}$$

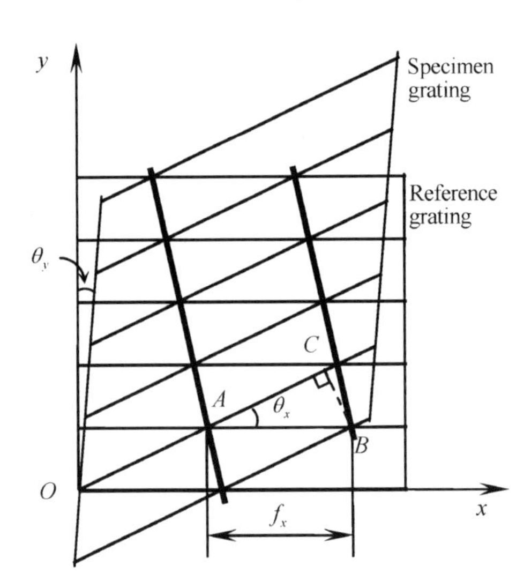

Fig. 4.5 Shearing strain measurement

In the second step, the grating lines of the reference and specimen gratings are parallel to the y axis. When the specimen grating is subjected to a shearing deformation, the angle θ_x makes the grating lines of the specimen grating move along the y direction and cannot produce moiré fringes, whereas the angle θ_y makes the grating lines of the specimen grating rotate and will cause moiré fringes. Using similar consideration, we have

$$\theta_y = \frac{p}{f_y} \tag{4.9}$$

where f_y is the spacing, in the y direction (i.e. the direction of grating lines), of two adjacent moiré fringes. Using Eq. (4.8) and Eq. (4.9), the shearing strain being measured can be expressed as

$$\gamma_{xy} = \theta_x + \theta_y = \frac{p}{f_x} + \frac{p}{f_y} \tag{4.10}$$

3. Plane Strain Measurement

For plane strain, ε_x, ε_y and γ_{xy} are required to be determined simultaneously. Therefore, plane strain measurement also needs to be divided into two steps. In the first step, the grating lines of the reference and specimen gratings with the same pitch p are placed in the direction parallel to the x axis, as shown in Fig. 4.6. When the object is subjected to a plane deformation, the specimen grating will have the same deformation as the object and have the pitch p' after deformation. Since ε_x makes the grating lines of the specimen grating elongate or contract in the x direction and θ_y makes the grating lines of the specimen grating move along the x direction, ε_x and θ_y cannot cause moiré fringes. However, ε_y makes the pitch of the specimen grating enlarge or shrink and θ_x makes the grating lines of the specimen grating rotate, therefore ε_y and θ_x will cause moiré fringes. From Fig. 4.6, we can write

$$\sin\theta_x = \frac{BC}{AB} = \frac{p'}{f_x} \tag{4.11}$$

where f_x is the spacing in the x direction (i.e. the direction of grating lines) of two adjacent moiré fringes. Using $\sin\theta_x \approx \theta_x$ and $p' = (1+\varepsilon_y)p \approx p$, Eq. (4.11) can be simplified as

$$\theta_x = \frac{p}{f_x} \tag{4.12}$$

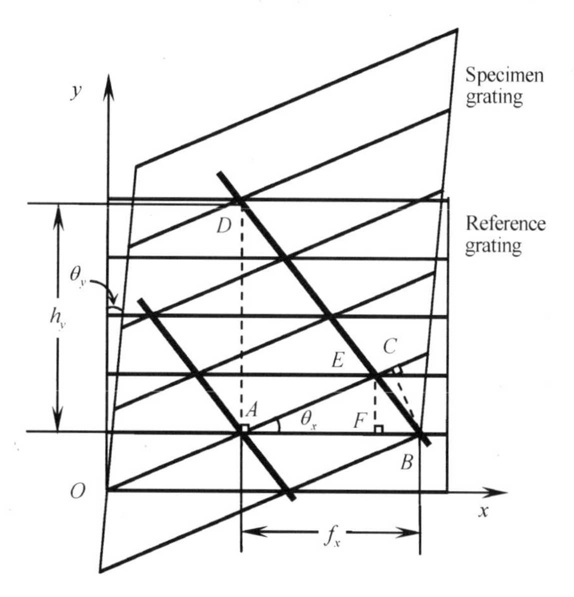

Fig. 4.6 Plane strain measurement

In addition, using the fact that $\triangle FBE$ and $\triangle ABD$ are two similar triangles, we obtain

$$FB = FE \cdot \frac{AB}{AD} = p \cdot \frac{f_x}{h_y} \tag{4.13}$$

where h_y is the spacing in the y direction (i.e. the direction perpendicular to the grating lines) of two adjacent moiré fringes. Since $\triangle AEF$ and $\triangle ABC$ are two similar triangles, we obtain

$$FB = AB - AF = AB - EF \cdot \frac{AC}{BC} = AB - EF \cdot \frac{\sqrt{AB^2 - BC^2}}{BC} = f_x - p \cdot \frac{\sqrt{f_x^2 - p'^2}}{p'} \tag{4.14}$$

Using $p' << f_x$, we have

$$FB = f_x - p \cdot \frac{f_x}{p'} = f_x - \frac{f_x}{1+\varepsilon_y} \tag{4.15}$$

From Eq. (4.13) and Eq. (4.15), we can obtain

$$\varepsilon_y = \frac{p}{h_y - p} \tag{4.16}$$

Using $p << h_y$, we have

$$\varepsilon_y = \frac{p}{h_y} \tag{4.17}$$

In the second step, the grating lines of the reference and specimen gratings are parallel to the y axis and assume that the pitch of the specimen grating is p before deformation and p' after deformation. ε_y makes the grating lines of the specimen grating elongate or contract in the y direction and θ_x makes the grating lines of the specimen grating move along the y direction, hence ε_y and θ_x cannot cause moiré fringes. However, ε_x makes the pitch of the specimen grating enlarge or shrink and θ_y makes the grating lines of the specimen grating rotate, therefore ε_x and θ_y will produce moiré fringes. Similarly, we can obtain

$$\theta_y = \frac{p}{f_y}, \quad \varepsilon_x = \frac{p}{h_x} \tag{4.18}$$

where f_y is the spacing in the y direction (i.e. the direction of grating lines) of two adjacent moiré fringes and h_x is the spacing in the x direction (i.e. the direction perpendicular to the grating lines). Using Eq. (4.12), Eq. (4.17) and Eq. (4.18), the plane strain being measured can be expressed as

$$\varepsilon_x = \frac{p}{h_x}, \quad \varepsilon_y = \frac{p}{h_y}, \quad \gamma_{xy} = \theta_x + \theta_y = \frac{p}{f_x} + \frac{p}{f_y} \tag{4.19}$$

It can be seen that when p, f_x, f_y, h_x, and h_y are known, ε_x, ε_y, and γ_{xy} can be determined by using Eq. (4.19).

4.1.3 Shadow Moiré for Out-of-Plane Displacement Measurement

The moiré method used for measuring the out-of-plane displacement of a deformed object is called shadow moiré. In shadow moiré, the specimen grating used is actually a shadow of the reference grating, which is formed on the surface of specimen.

Assume that a reference grating is placed in front of the object being measured, as shown in Fig. 4.7. A beam of light is used to illuminates the reference grating, and an observer (or a camera) receives the light that is scattered from the object surface. When the distance between the reference grating and the object surface is small, a shadow of the reference grating will be projected onto the object surface. This shadow, consisting of bright and dark lines, is itself a virtual grating. Therefore, the observer can see moiré fringes, which are formed due to interaction between the shadow and reference gratings.

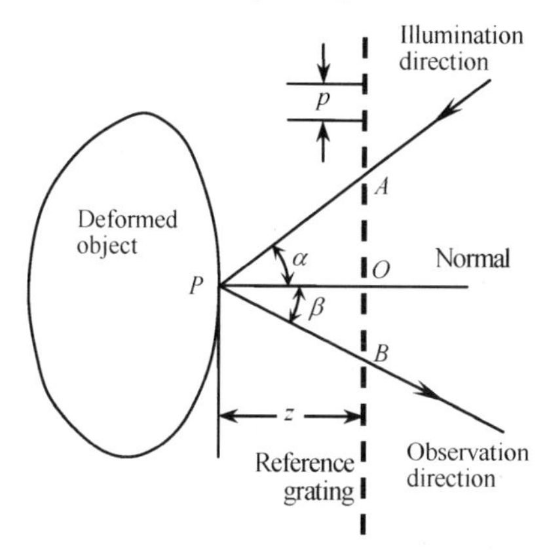

Fig. 4.7 Shadow moiré

Assume that the light passing through point A on the reference grating is projected onto point P on the object surface, where point P on the object surface is a shadow of point A on the reference grating. If point A is transparent, then point P is a bright spot. When we observe point P along the observation direction, as shown in Fig. 4.7, we can see that point P coincides with point B. If point B is transparent, then we can see a bright moiré fringe consisting of bright spots.

From Fig. 4.7, we have $AB = np$, where p is the pitch of the reference grating, n is the order of bright moiré fringes, i.e., $n = 0, 1, 2, \cdots$. Assuming that the vertical distance between the grating plane and point P on the object surface is equal to $OP = z$, then we obtain $AB = z(\tan\alpha + \tan\beta)$. Therefore, we have $np = z(\tan\alpha + \tan\beta)$, i.e.,

$$z = \frac{np}{\tan\alpha + \tan\beta} \tag{4.20}$$

where α is the included angle between the illumination direction and the normal of the reference grating, and β is the angle between the observation direction and the normal of the reference grating.

In shadow moiré, four different systems can be chosen for measuring the out-of-plane displacement: ①parallel illumination and parallel reception; ②diverging illumination and converging reception; ③ parallel illumination and converging reception; ④ diverging illumination and parallel reception. The following is a detailed analysis of the two commonly-used systems.

1. Parallel Illumination and Parallel Reception

As shown in Fig. 4.8, assuming that the included angle between the parallel incident light and the normal of the reference grating is α, and that the parallel reflected light is perpendicular to the reference grating, i.e. $\beta = 0$, then Eq. (4.20) can be rewritten by

$$z = \frac{np}{\tan\alpha} \tag{4.21}$$

Therefore, using the values of p and α, the distance z between the object point and the reference grating can be determined completely provided the order of moiré fringes is known.

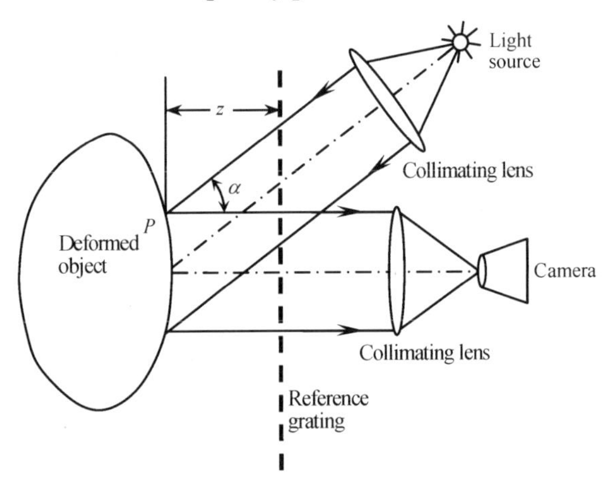

Fig. 4.8　Parallel illumination and parallel receiving

When two moiré patterns, one recorded before deformation and the other after deformation, are subjected to subtraction, we can obtain the out-of-plane displacement of the deformed object. Assuming that $z_0 = n_0 p / \tan\alpha$ and $z = np / \tan\alpha$ respectively correspond to the undeformed and deformed states, then we can express the out-of-plane displacement as

$$w = z - z_0 = \frac{np}{\tan\alpha} - \frac{n_0 p}{\tan\alpha} = \frac{(n - n_0)p}{\tan\alpha} \tag{4.22}$$

It can be seen from the above equation that, after the order of moiré fringes has been determined, the out-of-plane displacement at an arbitrary given point on the object surface can be determined based on Eq. (4.22).

2. Diverging Illumination and Converging Reception

The system corresponding to diverging illumination and converging reception is shown in

Fig. 4.9. In this system, it is assumed that the light source and the camera lens have the same vertical distance from the grating plane. From Fig. 4.9, we have $\tan\alpha = (L - x)/(D + z)$ and $\tan\beta = x/(D + z)$. Using Eq. (4.20), we obtain $z = np(D + z)/L$, i.e.,

$$z = \frac{npD}{L - np} \tag{4.23}$$

Using $np << L$, the above equation can be simplified as

$$z = \frac{npD}{L} \tag{4.24}$$

Therefore, the distance z between the object point and the reference grating can be determined completely provided the order of moiré fringes is known.

When two moiré patterns, which are respectively recorded before and after deformation, are subjected to subtraction, we can obtain the out-of-plane displacement of the deformed object as

$$w = z - z_0 = \frac{(n - n_0)pD}{L} \tag{4.25}$$

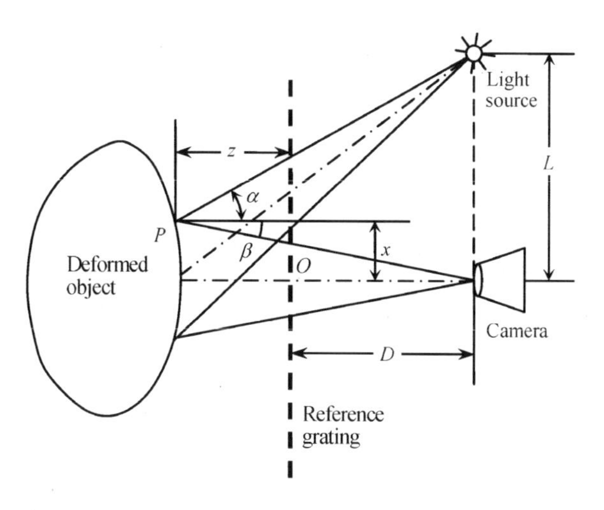

Fig. 4.9 Diverging illumination and converging receiving

4.1.4 Reflection Moiré for Slope Measurement

Reflection moiré can be used for measuring slopes (i.e. out-of-plane displacement derivatives). For a bent thin plate, the measurement of the out-of-plane displacement derivatives is more important than that of the out-of-plane displacement itself because the bending and twisting moments are directly related to the second-order derivatives of out-of-plane displacement (i.e., curvature and twist) instead of the out-of-plane displacement. It should be noted that the surface of specimen needs to be specular when reflection moiré is employed for measuring the out-of-plane displacement derivatives.

The measuring system used in reflection moiré is shown in Fig. 4.10. Assume that the light from point A on the grating plane is reflected from point P on the object surface and reaches point O on the image plane before deformation and that point O at the image plane will receive the light coming from point B on the grating plane after deformation. Two images of the grating respectively responding to the undeformed and deformed states of the plate, are recorded on a single photographic plate, and moiré fringes will be formed due to the fact that the recorded images have relative deformation.

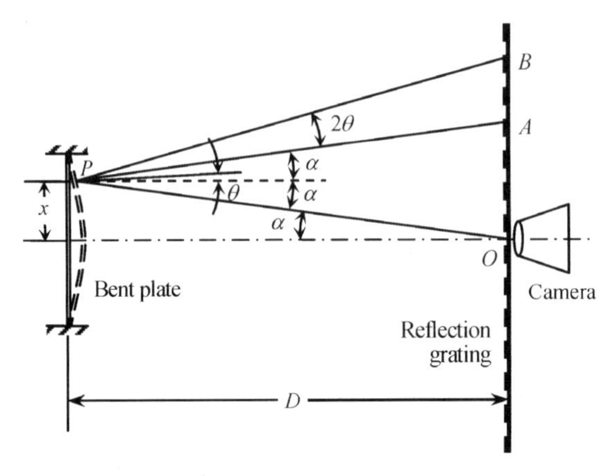

Fig. 4.10 Reflection Moiré

Assuming that the nth-order bright fringe is recorded, then we have $AB = np$, where p is the pitch of grating and n is the order of moiré fringes with $n = 0, 1, 2, \cdots$. If the vertical distance between the grating and the plate is equal to D, then we obtain $AB = D[\tan(\alpha + 2\theta) - \tan\alpha]$. Comparing the above two expressions, we obtain $np = D[\tan(\alpha + 2\theta) - \tan\alpha]$, i.e.,

$$\tan 2\theta = \frac{np}{D(1+\tan^2\alpha) + np} \tag{4.26}$$

Using $\tan\alpha = x/D << 1$ and $np << D$, the above expression can be simplified as

$$\tan 2\theta = \frac{np}{D} \tag{4.27}$$

For a small deformation, we have $\tan 2\theta \approx 2\partial w/\partial x$, or

$$\frac{\partial w}{\partial x} = \frac{np}{2D} \tag{4.28}$$

When the plate is rotated by $90°$, using a similar consideration, we can obtain

$$\frac{\partial w}{\partial y} = \frac{np}{2D} \tag{4.29}$$

It can be seen from both Eq. (4.28) and Eq. (4.29) that the slopes $\partial w/\partial x$ and $\partial w/\partial y$ can be determined provided the parameters p, D and n are known.

4.2 Moiré Interferometry

Since the diffraction grating with high density is used as the specimen grating in moiré interferometry, the measuring sensitivity of moiré interferometry is the same as that of holographic interferometry or speckle interferometry.

4.2.1 Real-Time Method for In-Plane Displacement Measurement

The measuring system used for in-plane displacement is shown in Fig. 4.11. When two beams of light, symmetrical to the optical axis, have the same incident angle equal to $\alpha = \arcsin(\lambda/p)$, the diffraction waves, O_1 and O_2 , can be obtained in the direction perpendicular to the surface of specimen. If the specimen grating affixed to the specimen surface is very regular, then the diffraction waves O_1 and O_2 can be regarded as two plane waves before the specimen is deformed, and can be expressed as

$$O_1 = a\exp(\mathrm{i}\varphi_1), \quad O_2 = a\exp(\mathrm{i}\varphi_2) \tag{4.30}$$

where a is the amplitude of each light wave, φ_1 and φ_2 are the phases of two light waves (for plane waves, φ_1 and φ_2 are both constants).

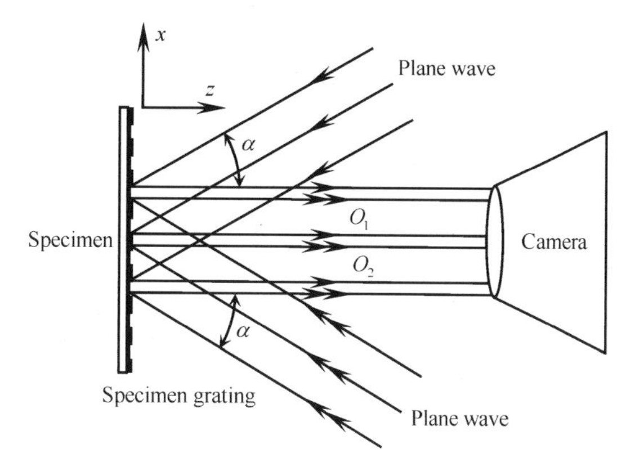

Fig. 4.11 In-plane displacement measurement system

When the specimen is deformed, the plane waves O_1 and O_2 will be changed into warped waves related to the deformation of specimen. These warped waves can be expressed as

$$O_1' = a\exp[\mathrm{i}(\varphi_1 + \delta_1)], \quad O_2' = a\exp[\mathrm{i}(\varphi_2 + \delta_2)] \tag{4.31}$$

where δ_1 and δ_2 are the phase change caused by the deformation of specimen and can be given by

$$\delta_1 = k[w(1+\cos\alpha) + u\sin\alpha], \quad \delta_2 = k[w(1+\cos\alpha) - u\sin\alpha] \tag{4.32}$$

where $w = w(x, y)$ and $u = u(x, y)$ are respectively the out-of-plane displacement and the in-plane displacement component along the x direction. The above warped waves passing through the imaging system will interfere with each other on the image plane and the intensity distribution recorded on the image plane can be expressed as

$$I=(O_1'+O_2')(O_1'+O_2')^*=2a^2[1+\cos(\varphi+\delta)] \tag{4.33}$$

where $\varphi=\varphi_1-\varphi_2$ is the initial phase difference between the plane waves O_1 and O_2, which is a constant equivalent to a uniform phase generated by translating the specimen. $\delta=\delta_1-\delta_2$ is the relative phase change of two warped waves after the deformation of specimen, which can be given, from Eq. (4.32), by

$$\delta=2ku\sin\alpha \tag{4.34}$$

In order to obtain the in-plane displacement component $v(x,y)$ along the y direction, both the specimen grating and the incident light waves should be rotated through $90°$. From the rotated system, we can obtain

$$\delta=2kv\sin\alpha \tag{4.35}$$

In order to obtain simultaneously two in-plane displacement components, an orthogonal grating needs to be duplicated onto the surface of specimen. This orthogonal grating can not only generate the diffraction halos along x and y directions, but also produce the diffraction halos along $+45°$ and $-45°$ direction. Therefore, all the in-plane displacement components along the x, y, $+45°$, and $-45°$ directions can be obtained simultaneously by using an orthogonal grating.

4.2.2 Differential Load Method for In-Plane Displacement Measurement

It is generally difficult to obtain absolutely accurate plane waves of diffraction, thus the phase difference $\varphi=\varphi_1-\varphi_2$ is not constant. When φ is not constant, the differential load method needs to be used for measuring in-plane displacement components. In order to eliminate the influence of φ on displacement fringes, an optical wedge is required to be appended to the optical path O_1 (or O_2), and the two diffractive optical waves before loading are respectively given by

$$O_1=a\exp[\mathrm{i}(\varphi_1+f)],\quad O_2=a\exp(\mathrm{i}\varphi_2) \tag{4.36}$$

where $f=f(x,y)$ is the phase distribution of the optical wedge. The corresponding intensity distribution is given by

$$I_1=(O_1+O_2)(O_1+O_2)^*=2a^2[1+\cos(\varphi+f)] \tag{4.37}$$

where $\varphi=\varphi_1-\varphi_2$.

The diffraction light waves after loading can be expressed as

$$O_1'=a\exp[\mathrm{i}(\varphi_1+\delta_1+f)],\quad O_2'=a\exp[\mathrm{i}(\varphi_2+\delta_2)] \tag{4.38}$$

where $\delta_1=k[w(1+\cos\alpha)+u\sin\alpha]$ and $\delta_2=k[w(1+\cos\alpha)-u\sin\alpha]$. The corresponding intensity distribution is

$$I_2 = (O_1' + O_2')(O_1' + O_2')^* = 2a^2[1 + \cos(\varphi + \delta + f)] \tag{4.39}$$

where $\delta = \delta_1 - \delta_2 = 2ku\sin\alpha$.

The total intensity distribution after two series of exposure can be written by

$$I = I_1 + I_2 = 4a^2\left[1 + \cos\left(\varphi + \frac{1}{2}\delta + f\right)\cos\left(\frac{1}{2}\delta\right)\right] \tag{4.40}$$

When the double-exposed photographic plate after development and fixation is placed into a filtering system, dark fringes can be obtained in accordance with the following condition:

$$\cos\left(\frac{1}{2}\delta\right) = 0 \tag{4.41}$$

i.e.,

$$\delta = (2n+1)\pi \quad (n = 0, \pm 1, \pm 2, \cdots) \tag{4.42}$$

Substituting Eq. (4.34) into Eq. (4.42), we have

$$u = \frac{(2n+1)\pi}{2k\sin\alpha} = \frac{(2n+1)\lambda}{4\sin\alpha} \quad (n = 0, \pm 1, \pm 2, \cdots) \tag{4.43}$$

Chapter 5 Phase-Shifting Interferometry and Phase Unwrapping

In optical measurement mechanics, the quantities to be measured (e.g. displacement, strain, etc.) are directly related to the phase information contained in interference fringes; thus it is quite important to extract the phase information from these interference fringes. The optical measurement techniques, such as holographic interferometry, speckle interferometry, and moiré interferometry, etc., record interference fringes formed due to mutual interference of coherent light beams. The interference fringes denote the contour lines of phase, i.e., various points on each fringe centerline have the same phase value, and arbitrary centerlines of adjacency have the same phase difference.

The traditional phase detection method needs to determine the location and order of each fringe so as to obtain the phase distribution on the interference fringe pattern. However, the traditional method often causes large error of measurement for the reason that the maximum brightness is not located at the centerline of fringes and the phase of any point between two adjacent fringes can only be obtained by interpolation. In order to eliminate deficiency of the traditional method, a variety of phase detection methods have been proposed.

In optical measurement mechanics, the commonly-used phase detection method is phase-shifting interferometry. This method does not need to determine location and order of each fringe and can obtain directly phases of various points on an interference fringe pattern.

5.1 Phase-Shifting Interferometry

Phase-shifting interferometry refers to a technique used for determining the distribution of phase by recording three or more interference fringe patterns with known phase-shifting values between arbitrary two interference fringe patterns or by recording one interference fringe pattern with known phase-shifting values between arbitrary two adjacent points on the recorded interference fringe pattern.

Phase-shifting interferometry is usually divided into temporal phase-shifting interferometry and spatial phase-shifting interferometry.

5.1.1 Temporal Phase-Shifting Interferometry

1. Temporal phase-shifting Principle

The intensity distribution of an interference fringe patten obtained in optical measurement mechanics can be expressed as

$$I(x,y)=I_0(x,y)[1+V(x,y)\cos\delta(x,y)] \tag{5.1}$$

where $I_0(x,y)$ is the background intensity, $V(x,y)$ the fringe visibility (or modulation), and $\delta(x,y)$ the phase to be determined.

When the above interference fringe pattern is sampled and quantized into a digital image by using a CCD (charge coupled device) or CMOS (complementary metal oxide semiconductor) camera, the intensity distribution of interference fringe pattern will be degraded by electronic noise, speckle noise, etc. Therefore, when all these factors are considered, the intensity distribution of interference fringe pattern can be rewritten by

$$I(x,y)=A(x,y)+B(x,y)\cos\delta(x,y) \tag{5.2}$$

where $A(x,y)$ and $B(x,y)$ are, respectively, the background intensity and modulation intensity. Eq. (5.2) involves three unknown quantities $A(x,y)$, $B(x,y)$ and $\delta(x,y)$. Therefore, at least three independent equations are required to determine the phase $\delta(x,y)$ at each point on an interference fringe pattern.

Assuming that the phase-shifting value corresponding to the nth interference fringe pattern is denoted by α_n, then the intensity distribution of the nth interference fringe pattern can be given by

$$I_n(x,y)=A(x,y)+B(x,y)\cos[\delta(x,y)+\alpha_n] \quad (n=1,2,\cdots,N;N\geqslant 3) \tag{5.3}$$

where $I_n(x,y)$ and α_n are two known quantities, while $A(x,y)$, $B(x,y)$ and $\delta(x,y)$ are three unknown quantities. Hence, using at least different phase-shifting values, the distribution of phase $\delta(x,y)$ can be completely determined.

2. Temporal Phase-Shifting Algorithms

Temporal phase-shifting interferometry mainly includes the three-step algorithm, four-step algorithm, Carré algorithm, etc. These algorithms have been widely used in optical measurement mechanics.

1) Three-Step Algorithm

Assuming three phase-shifting values are α_1, α_2 and α_3 respectively, then the intensity distributions of three phase-shifted interference fringe patterns can be expressed as

$$\begin{aligned}
I_1(x,y)&=A(x,y)+B(x,y)\cos[\delta(x,y)+\alpha_1]\\
I_2(x,y)&=A(x,y)+B(x,y)\cos[\delta(x,y)+\alpha_2]\\
I_3(x,y)&=A(x,y)+B(x,y)\cos[\delta(x,y)+\alpha_3]
\end{aligned} \tag{5.4}$$

Solving the above three equations for $\delta(x,y)$, the distribution of phase can be expressed as

$$\frac{\tan\delta(x,y)(\sin\alpha_2-\sin\alpha_3)-(\cos\alpha_2-\cos\alpha_3)}{\tan\delta(x,y)(2\sin\alpha_1-\sin\alpha_2-\sin\alpha_3)-(2\cos\alpha_1-\cos\alpha_2-\cos\alpha_3)}=\frac{I_2(x,y)-I_3(x,y)}{2I_1(x,y)-I_2(x,y)-I_3(x,y)} \tag{5.5}$$

Eq. (5.5) is a general expression for the three-step algorithm. Some particular cases of the above equation are outlined as follows:

(1) If three phase-shifting values are 0, $\pi/3$ and $2\pi/3$ (i.e. the phase-shifting increment between any two adjacent interference fringe patterns is $\pi/3$), the distribution of phase is equal to

$$\delta(x,y)=\arctan\frac{2I_1(x,y)-3I_2(x,y)+I_3(x,y)}{\sqrt{3}[I_2(x,y)-I_3(x,y)]} \tag{5.6}$$

(2) If three phase-shifting values are 0, $\pi/2$ and π (the phase-shifting increment is $\pi/2$), the distribution of phase is equal to

$$\delta(x,y)=\arctan\frac{I_1(x,y)-2I_2(x,y)+I_3(x,y)}{I_1(x,y)-I_3(x,y)} \tag{5.7}$$

(3) If three phase-shifting values are 0, $2\pi/3$ and $4\pi/3$ (the phase-shifting increment is $2\pi/3$), the distribution of phase is equal to

$$\delta(x,y)=\arctan\frac{\sqrt{3}[I_3(x,y)-I_2(x,y)]}{2I_1(x,y)-I_2(x,y)-I_3(x,y)} \tag{5.8}$$

2) Four-Step Algorithm

Assuming four phase-shifting values are α_1, α_2, α_3 and α_4 respectively, then the intensity distributions of four phase-shifted interference fringe patterns can be expressed as

$$\begin{aligned}
I_1(x,y)&=A(x,y)+B(x,y)\cos[\delta(x,y)+\alpha_1]\\
I_2(x,y)&=A(x,y)+B(x,y)\cos[\delta(x,y)+\alpha_2]\\
I_3(x,y)&=A(x,y)+B(x,y)\cos[\delta(x,y)+\alpha_3]\\
I_4(x,y)&=A(x,y)+B(x,y)\cos[\delta(x,y)+\alpha_4]
\end{aligned} \tag{5.9}$$

Solving these equations for $\delta(x,y)$, the distribution of phase can be expressed as

$$\frac{\tan\delta(x,y)(\sin\alpha_1-\sin\alpha_3)-(\cos\alpha_1-\cos\alpha_3)}{\tan\delta(x,y)(\sin\alpha_2-\sin\alpha_4)-(\cos\alpha_2-\cos\alpha_4)}=\frac{I_1(x,y)-I_3(x,y)}{I_2(x,y)-I_4(x,y)} \tag{5.10}$$

Eq. (5.10) is also a general expression for the four-step algorithm. Some particular cases include:

(1) If four phase-shifting values are $\pi/4$, $3\pi/4$, $5\pi/4$ and $7\pi/4$ (the phase-shifting increment is $\pi/2$), the distribution of phase is equal to

$$\delta(x,y)=\arctan\frac{[I_2(x,y)-I_4(x,y)]+[I_1(x,y)-I_3(x,y)]}{[I_2(x,y)-I_4(x,y)]-[I_1(x,y)-I_3(x,y)]} \tag{5.11}$$

(2) If four phase-shifting values are 0, $\pi/3$, $2\pi/3$ and π (the phase-shifting increment is $\pi/3$), the distribution of phase is equal to

$$\delta(x,y)=\arctan\frac{I_1(x,y)-I_2(x,y)-I_3(x,y)+I_4(x,y)}{\sqrt{3}[I_2(x,y)-I_3(x,y)]} \tag{5.12}$$

(3) If four phase-shifting values are 0, $\pi/2$, π and $3\pi/2$ (the phase-shifting increment is $\pi/2$), the distribution of phase is equal to

$$\delta(x,y) = \arctan\frac{I_4(x,y) - I_2(x,y)}{I_1(x,y) - I_3(x,y)} \tag{5.13}$$

3) Carré Algorithm

The Carré algorithm is one of the commonly-used algorithms in optical measurement mechanics. If four phase-shifting values are -3α, $-\alpha$, α and 3α (the phase-shifting increment is 2α, while the value of α is unknown), the intensity distributions of four phase-shifted interference fringe patterns can be expressed as

$$\begin{aligned}
I_1(x,y) &= A(x,y) + B(x,y)\cos[\delta(x,y) - 3\alpha] \\
I_2(x,y) &= A(x,y) + B(x,y)\cos[\delta(x,y) - \alpha] \\
I_3(x,y) &= A(x,y) + B(x,y)\cos[\delta(x,y) + \alpha] \\
I_4(x,y) &= A(x,y) + B(x,y)\cos[\delta(x,y) + 3\alpha]
\end{aligned} \tag{5.14}$$

Solving the above equations, the distribution of phase can be expressed as

$$\delta(x,y) = \arctan\left\{\tan\beta\frac{[I_2(x,y) - I_3(x,y)] + [I_1(x,y) - I_4(x,y)]}{[I_2(x,y) + I_3(x,y)] - [I_1(x,y) + I_4(x,y)]}\right\} \tag{5.15}$$

where β is equal to

$$\tan^2\beta = \frac{3[I_2(x,y) - I_3(x,y)] - [I_1(x,y) - I_4(x,y)]}{[I_2(x,y) - I_3(x,y)] + [I_1(x,y) - I_4(x,y)]} \tag{5.16}$$

3. Temporal Phase-Shifting Experiment

Three interference fringe patterns obtained by using the three-step temporal phase-shifting algorithm are shown in Fig. 5.1. The phase-shifting values are 0, $\pi/2$ and π, respectively corresponding to Fig. 5.1(a), Fig. 5.1(b) and Fig. 5.1(c).

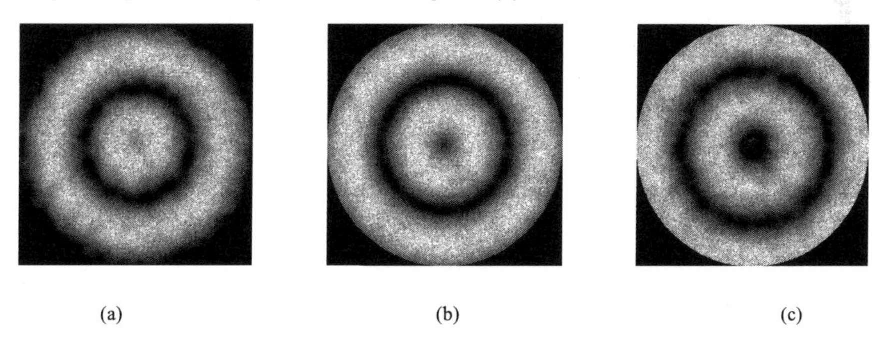

(a) (b) (c)

Fig. 5.1 Interference fringe patterns

Using the above interference fringe patterns, the wrapped phase map obtained based on the three-step algorithm is shown in Fig. 5.2. The wrapped phase is distributed in the range of $-\pi/2 \sim \pi/2$.

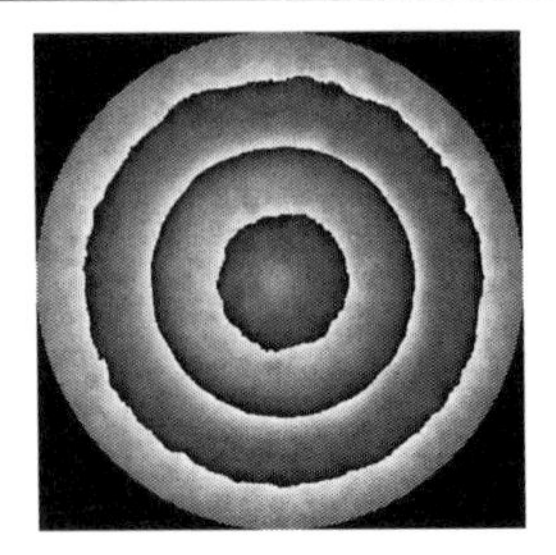

Fig. 5.2　Wrapped phase map

5.1.2　Spatial Phase-Shifting Interferometry

1. Spatial Phase-Shifting Principle

The commonly-used technique in spatial phase-shifting interferometry is to introduce a spatial carrier. Hence this technique is often called the spatial carrier method.

The intensity distribution of an interference fringe pattern after introducing a spatial carrier can be given by

$$I(x,y)=A(x,y)+B(x,y)\cos[\delta(x,y)+2\pi fx] \tag{5.17}$$

where $f=f(x)$ is a linear spatial carrier along the x direction (i.e. the carrier direction), and $\delta(x,y)$ is the distribution of phase, which is to be determined.

When the interference fringe pattern recorded by a CCCD camera is stored in the form of a digital image, the intensity value at pixel (i,j) is equal to

$$I(x_i,y_j)=A(x_i,y_j)+B(x_i,y_j)\cos[\delta(x_i,y_j)+2\pi fx_i] \tag{5.18}$$

where $i=1,2,\cdots,M;\ j=1,2,\cdots,N$ with $M\times N$ being the total pixels of the CCD used.

2. Spatial Phase-Shifting Algorithms

1) Three-Step Algorithm

Assuming that three adjacent pixels (i,j), $(i+1,j)$ and $(i+2,j)$ have the same background intensity, modulation intensity, and phase to be measured, the intensity distributions at these pixels can be expressed as

$$\begin{aligned}
I(x_i,y_j)&=A(x_i,y_j)+B(x_i,y_j)\cos[\delta(x_i,y_j)+2\pi fx_i]\\
I(x_{i+1},y_j)&=A(x_i,y_j)+B(x_i,y_j)\cos[\delta(x_i,y_j)+2\pi f(x_i+\Delta x)]\\
I(x_{i+2},y_j)&=A(x_i,y_j)+B(x_i,y_j)\cos[\delta(x_i,y_j)+2\pi f(x_i+2\Delta x)]
\end{aligned} \tag{5.19}$$

where Δx is the pixel width of the CCD camera along the x direction, and $2\pi fx_i$ is the carrier phase at pixel (i,j) with $i=1,2,\cdots,M-2;\ j=1,2,\cdots,N$.

When the three-step algorithm is utilized, the carrier fringe width is chosen to be equal to the width of three pixels of the CCD camera so that the phase-shifting increment is equal to

$2\pi/3$ in the x direction (i.e. $2\pi f\Delta x = 2\pi/3$). Therefore, the above three equations can be rewritten by

$$\begin{aligned} I(x_i, y_j) &= A(x_i, y_j) + B(x_i, y_j)\cos[\delta(x_i, y_j) + 2\pi f x_i] \\ I(x_{i+1}, y_j) &= A(x_i, y_j) + B(x_i, y_j)\cos\left[\delta(x_i, y_j) + 2\pi f x_i + \frac{2}{3}\pi\right] \\ I(x_{i+2}, y_j) &= A(x_i, y_j) + B(x_i, y_j)\cos\left[\delta(x_i, y_j) + 2\pi f x_i + \frac{4}{3}\pi\right] \end{aligned} \tag{5.20}$$

Solving the above equations, the wrapped phase can be expressed as

$$\delta(x_i, y_j) + 2\pi f x_i = \arctan\left\{\frac{\sqrt{3}[I(x_{i+2}, y_j) - I(x_{i+1}, y_j)]}{2I(x_i, y_j) - I(x_{i+1}, y_j) - I(x_{i+2}, y_j)}\right\} \tag{5.21}$$

2) Four-Step Algorithm

Assuming that four adjacent pixels (i, j), $(i+1, j)$, $(i+2, j)$ and $(i+3, j)$ have the same background intensity, modulation intensity, and phase to be measured, the intensity distributions at pixels (i, j), $(i+1, j)$, $(i+2, j)$ and $(i+3, j)$ can be expressed as

$$\begin{aligned} I(x_i, y_j) &= A(x_i, y_j) + B(x_i, y_j)\cos[\delta(x_i, y_j) + 2\pi f x_i] \\ I(x_{i+1}, y_j) &= A(x_i, y_j) + B(x_i, y_j)\cos[\delta(x_i, y_j) + 2\pi f(x_i + \Delta x)] \\ I(x_{i+2}, y_j) &= A(x_i, y_j) + B(x_i, y_j)\cos[\delta(x_i, y_j) + 2\pi f(x_i + 2\Delta x)] \\ I(x_{i+3}, y_j) &= A(x_i, y_j) + B(x_i, y_j)\cos[\delta(x_i, y_j) + 2\pi f(x_i + 3\Delta x)] \end{aligned} \tag{5.22}$$

where $i = 1, 2, \cdots, M-3;\ j = 1, 2, \cdots, N$.

If the four-step algorithm is employed, the carrier fringe width should be equal to the width of four pixels of the CCD camera so that the phase-shifting increment is $\pi/2$ (i.e. $2\pi f\Delta x = \pi/2$). Hence the above four equations can be rewritten by

$$\begin{aligned} I(x_i, y_j) &= A(x_i, y_j) + B(x_i, y_j)\cos[\delta(x_i, y_j) + 2\pi f x_i] \\ I(x_{i+1}, y_j) &= A(x_i, y_j) + B(x_i, y_j)\cos\left[\delta(x_i, y_j) + 2\pi f x_i + \frac{1}{2}\pi\right] \\ I(x_{i+2}, y_j) &= A(x_i, y_j) + B(x_i, y_j)\cos[\delta(x_i, y_j) + 2\pi f x_i + \pi] \\ I(x_{i+3}, y_j) &= A(x_i, y_j) + B(x_i, y_j)\cos\left[\delta(x_i, y_j) + 2\pi f x_i + \frac{3}{2}\pi\right] \end{aligned} \tag{5.23}$$

Solving these equations, the wrapped phase can be given by

$$\delta(x_i, y_j) + 2\pi f x_i = \arctan\left[\frac{I(x_{i+3}, y_j) - I(x_{i+1}, y_j)}{I(x_i, y_j) - I(x_{i+2}, y_j)}\right] \tag{5.24}$$

3. Spatial Phase-Shifting Experiment

The experimental result obtained by using the spatial carrier method is shown in Fig. 5.3. Fig. 5.3(a) is the modulated interference fringe pattern, Fig. 5.3(b) is the wrapped phase distribution.

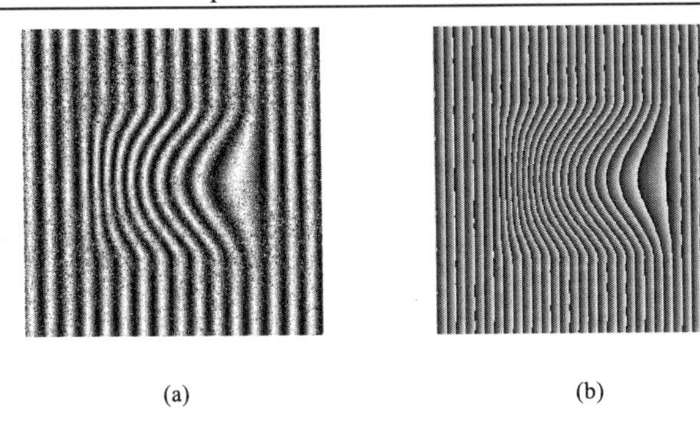

(a) (b)

Fig. 5.3 Experimental result obtained using spatial carrier

5.2 Phase Unwrapping

1. Phase Unwrapping Principle

The distribution of phase obtained from phase-shifting interferometry can expressed as

$$\delta(x,y)=\arctan\frac{S(x,y)}{C(x,y)} \tag{5.25}$$

It can be seen from Eq. (5.25) that $\delta(x,y)$ is always a wrapped phase with its phase values falling in the range of $-\pi/2\sim\pi/2$. According to the signs of $S(x,y)$ and $C(x,y)$, the above wrapped phase can be expanded into the range of $0\sim2\pi$ by using the following transformation:

$$\delta(x,y)=\begin{cases}\delta(x,y) & (S(x,y)\geqslant 0,\ C(x,y)>0)\\ \dfrac{1}{2}\pi & (S(x,y)>0,\ C(x,y)=0)\\ \delta(x,y)+\pi & (C(x,y)<0)\\ \dfrac{3}{2}\pi & (S(x,y)<0,\ C(x,y)=0)\\ \delta(x,y)+2\pi & (S(x,y)<0,\ C(x,y)>0)\end{cases} \tag{5.26}$$

After the phase is expanded, the distribution of phase has been changed from $-\pi/2\sim\pi/2$ into $0\sim2\pi$.

The above expanded phase is still a wrapped phase having its phase values in the range of $0\sim2\pi$. To obtain a continuous phase distribution, the above expanded phase needs to be unwrapped by using the 2D phase unwrapping algorithm. If the phase difference between two adjacent pixels is equal to or larger than π, then a 2π ambiguity can be removed by adding or subtracting 2π until phase difference between these two adjacent pixels is smaller than π. Therefore, the distribution of unwrapped phase can be given by

$$\delta_{\mathrm{u}}(x,y)=\delta(x,y)+2\pi n(x,y) \tag{5.27}$$

where $n(x,y)$ is an integer.

2. Phase Unwrapping Experiments

Expt. 1 The phase maps obtained by using the temporal phase-shifting algorithm are shown in Fig. 5.4. Fig. 5.4(a) and Fig. 5.4(b) are the wrapped phase maps, which have the phase values of $-\pi/2 \sim \pi/2$ and of $0 \sim 2\pi$, respectively, and Fig. 5.4(c) is the unwrapped phase map.

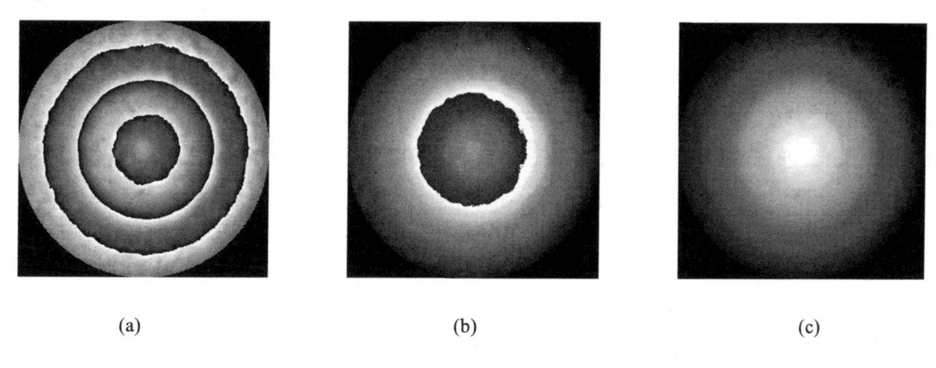

(a) (b) (c)

Fig. 5.4 Phase maps obtained using temporal phase-shifting algorithm

Expt. 2 The phase maps obtained by using the spatial carrier algorithm are shown in Fig. 5.5. Fig. 5.5(a) is the modulated wrapped phase map, Fig. 5.5(b) is the modulated unwrapped phase map, and Fig. 5.5(c) is the unwrapped phase map without carrier.

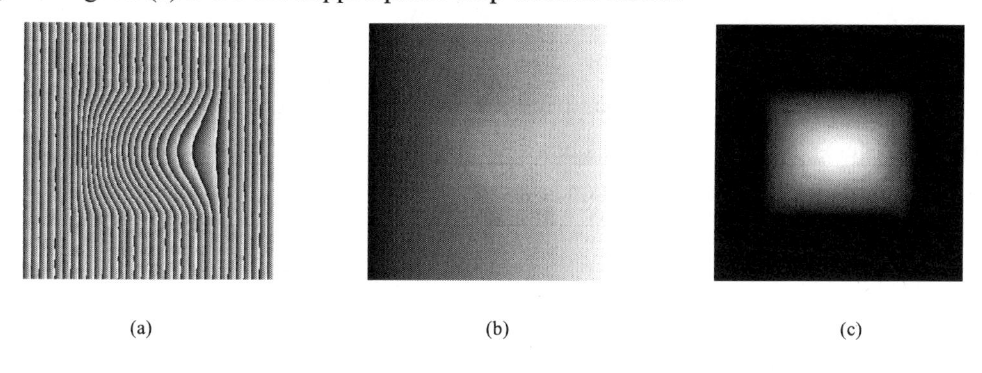

(a) (b) (c)

Fig. 5.5 Phase maps obtained using spatial carrier algorithm

Chapter 6 Discrete Transformation and Low-Pass Filtering

Interference fringes obtained in optical measurement mechanics, especially in digital speckle interferometry, usually involve heavy noise. Thus it is required to choose a proper filtering method to reduce or remove the noise from these interference fringes prior to calculating the phase distribution of an interference fringe pattern. Since the low-pass filtering methods can used to reduce or remove the noise of the interference fringes, they have been widely used for denoising the interference fringes obtained in optical measurement mechanics.

The denoising of interference fringes can be performed in either the spatial or the frequency domain. If this is done in the frequency domain, then an interference fringe pattern is required to be transformed from the spatial domain into the frequency domain.

6.1 Discrete Transformation

The commonly-used methods of discrete transformation mainly include the discrete Fourier transform, the discrete cosine transform, etc.

6.1.1 Discrete Fourier Transform

The discrete Fourier transform (DFT) converts a finite sequence of equally spaced samples of a function into the list of coefficients of a finite combination of complex sinusoids, ordered by their frequencies, that has the same sample values. It can convert the sampled function from the space domain to the frequency domain.

The discrete Fourier transform is designed for processing a real- or complex-valued signal, and it always produce a complex-valued spectrum even in the case where the original signal was strictly real-valued. The reason is that neither the real nor the imaginary part of this Fourier spectrum alone is sufficient to represent (i.e., reconstruct) the original signal completely. In other words, the corresponding cosine (for the real part) or sine functions (for the imaginary part) alone do not constitute a complete set of basis functions. A real-valued signal always has a centosymmetric Fourier spectrum, so only one half of the spectral coefficients need to be computed without losing any signal information.

The discrete Fourier transform is the most important discrete transform, used to perform Fourier analysis in many practical applications. In digital signal processing, the function is any quantity that varies over time, such as the pressure of a sound wave, a radio signal, etc., sampled over a finite time interval. In digital image processing, the samples can be the values of pixels along a row or column of a digital image. The discrete Fourier transform is also used to perform other operations such as convolution, correlation, etc.

1. Discrete Fourier Transform Principle

1) One-Dimensional Discrete Fourier Transform

For a discrete signal $f(m)$ of length M ($m=0,1,\cdots,M-1$), the discrete Fourier transform of this signal is defined as

$$F(p)=\sum_{m=0}^{M-1} f(m)\exp\left(-\mathrm{i}2\pi\frac{mp}{M}\right) \quad (p=0,1,\cdots,M-1) \tag{6.1}$$

where p is the pixel coordinate in the frequency-domain, $\exp\left(-\mathrm{i}2\pi\frac{mp}{M}\right)$ is the (forward) transform kernel.

The inverse discrete Fourier transform is defined as

$$f(m)=\frac{1}{M}\sum_{p=0}^{M-1} F(p)\exp\left(\mathrm{i}2\pi\frac{mp}{M}\right) \quad (m=0,1,\cdots,M-1) \tag{6.2}$$

where $\exp\left(\mathrm{i}2\pi\frac{mp}{M}\right)$ is the inverse transform kernel.

2) Two-Dimensional Discrete Fourier Transform

For a two-dimensional digital image $f(m,n)$ of size $M\times N$, the two-dimensional discrete Fourier transform is defined as

$$F(p,q)=\sum_{m=0}^{M-1}\sum_{n=0}^{N-1} f(m,n)\exp\left[-\mathrm{i}2\pi\left(\frac{mp}{M}+\frac{nq}{N}\right)\right] \quad (p=0,1,\cdots,M-1;q=0,1,\cdots,N-1) \tag{6.3}$$

where (p,q) are the pixel coordinates in the frequency-domain. As we see, the resulting Fourier transform is again a two-dimensional function of size $M\times N$.

Similarly, the two-dimensional inverse discrete Fourier transform is defined as

$$f(m,n)=\frac{1}{MN}\sum_{p=0}^{M-1}\sum_{q=0}^{N-1} F(p,q)\exp\left[\mathrm{i}2\pi\left(\frac{mp}{M}+\frac{nq}{N}\right)\right] \quad (m=0,1,\cdots,M-1;n=0,1,\cdots,N-1) \tag{6.4}$$

2. Fast Fourier Transform Principle

Since the discrete Fourier transform deals with a finite amount of data, it can be implemented in computers by numerical algorithms or even dedicated hardware. However, the discrete Fourier transform with a million points is common in many applications. Therefore, modern signal and image processing would be impossible without an efficient method for computing the discrete Fourier transform. Fortunately, fast Fourier transform (FFT) algorithms can be chosen for computing the discrete Fourier transform. These fast algorithms rearrange the sequence of computations in such a way that intermediate results are only computed once and optimally reused many times. Since its invention, the fast Fourier transform has therefore become an indispensable tool in almost any application of signal and image analysis.

The execution time of the fast Fourier transform depends on the transform length. It is fastest when the transform length is a power of two, and almost as fast when the transform length has only small prime factors. It is typically slower for transform lengths that are prime or have large prime factors. Time differences, however, are reduced to insignificance by modern fast Fourier transform algorithms such as those used in MATLAB. Adjusting the transform length is usually unnecessary in practice. The fast Fourier transform generally reduces the time complexity of the computation. The benefits are substantial, in particular for longer signals or greater images.

1) One-Dimensional Fast Fourier Transform

Assuming that the transform kernel is denoted by

$$W_M^{mp} = \exp\left(-\mathrm{i}2\pi\frac{mp}{M}\right) \tag{6.5}$$

then the one-dimensional discrete Fourier transform can be expressed as

$$F(p) = \sum_{m=0}^{M-1} f(m)W_M^{mp} \quad (p = 0,1,\cdots,M-1) \tag{6.6}$$

Eq. (6.6) shows that, for a data vector of length M, direct calculation of the discrete Fourier transform requires M^2 multiplications and $M(M-1)$ additions, i.e. a total of $M(2M-1)$ floating-point operations. To compute a million-point discrete Fourier transform, a computer capable of doing one multiplication and addition every microsecond requires a million seconds, or about 11.57 days. When using the fast Fourier transform algorithm, the calculation of Eq. (6.6) will be divided into two steps.

Step 1: considering $F(p)$ within the range of $(0,1,\cdots,M/2-1)$, then Eq. (6.6) can be given by

$$F(p) = \sum_{m=0}^{M/2-1} f(2m)W_M^{2mp} + \sum_{m=0}^{M/2-1} f(2m+1)W_M^{(2m+1)p} \quad (p = 0,1,\cdots,M/2-1) \tag{6.7}$$

Using $W_M^{2mp} = W_{M/2}^{mp}$, we obtain

$$F(p) = \sum_{m=0}^{M/2-1} f(2m)W_{M/2}^{mp} + W_M^p \sum_{m=0}^{M/2-1} f(2m+1)W_{M/2}^{mp} \quad (p = 0,1,\cdots,M/2-1) \tag{6.8}$$

Assuming $F_\mathrm{e}(p) = \sum_{m=0}^{M/2-1} f(2m)W_{M/2}^{mp}$ and $F_\mathrm{o}(p) = \sum_{m=0}^{M/2-1} f(2m+1)W_{M/2}^{mp}$, then Eq.(6.8) can be expressed as

$$F(p) = F_\mathrm{e}(p) + W_M^p F_\mathrm{o}(p) \quad (p = 0,1,\cdots,M/2-1) \tag{6.9}$$

where $F_\mathrm{e}(p)$ and $F_\mathrm{o}(p)$ are respectively the even and odd terms.

Step 2: considering $F(p)$ within the range of $(M/2, M/2+1,\cdots,M-1)$, Eq. (6.6) can be rewritten by

$$F(p+M/2) = \sum_{m=0}^{M/2-1} f(2m)W_M^{2m(p+M/2)} + \sum_{m=0}^{M/2-1} f(2m+1)W_M^{(2m+1)(p+M/2)} \quad (p=0,1,\cdots,M/2-1) \tag{6.10}$$

Using $W_M^{2mp} = W_{M/2}^{mp}$, $W_M^{mM} = 1$, and $W_M^{M/2} = -1$, we have

$$F(p+M/2)=\sum_{m=0}^{M/2-1} f(2m)W_{M/2}^{mp}-W_M^p\sum_{m=0}^{M/2-1} f(2m+1)W_{M/2}^{mp}=F_e(p)-W_M^pF_o(p) \quad (p=0,1,\cdots,M/2-1) \tag{6.11}$$

2) Two-Dimensional Fast Fourier Transform

The two-dimensional fast Fourier transform can be given by

$$F(p,q)=\sum_{m=0}^{M-1}\sum_{n=0}^{N-1} f(m,n)W_M^{mp}W_N^{nq} \quad (p=0,1,\cdots,M-1; q=0,1,\cdots,N-1) \tag{6.12}$$

where $W_M^{mp}=\exp\left(-\mathrm{i}2\pi\frac{mp}{M}\right)$ and $W_N^{nq}=\exp\left(-\mathrm{i}2\pi\frac{nq}{N}\right)$. Eq. (6.12) can also be written by

$$F(p,q)=\sum_{m=0}^{M-1}[\sum_{n=0}^{N-1} f(m,n)W_N^{nq}]W_M^{mp} \quad (p=0,1,\cdots,M-1; q=0,1,\cdots,N-1) \tag{6.13}$$

or

$$F(p,q)=\sum_{n=0}^{N-1}[\sum_{m=0}^{M-1} f(m,n)W_M^{mp}]W_N^{nq} \quad (p=0,1,\cdots,M-1; q=0,1,\cdots,N-1) \tag{6.14}$$

Eq. (6.14) shows that one two-dimensional fast Fourier transform can be converted into two one-dimensional fast Fourier transforms.

3. Discrete Fourier Transform Experiment

The discrete Fourier transform can transform a digital image between the space domain and the frequency domain.

The experimental results of the discrete Fourier transform are shown in Fig. 6.1. Fig. 6.1(a) and Fig. 6.1(b) are the digital specklegrams, respectively corresponding to the undeformed and deformed states; Fig. 6.1(c) is the Fourier spectrum; Fig. 6.1(d) is the designed ideal band-pass filter; Fig. 6.1(e) is the Fourier spectrum after filtering; Fig. 6.1(f) is the contour fringes of out-of-plane displacement derivative.

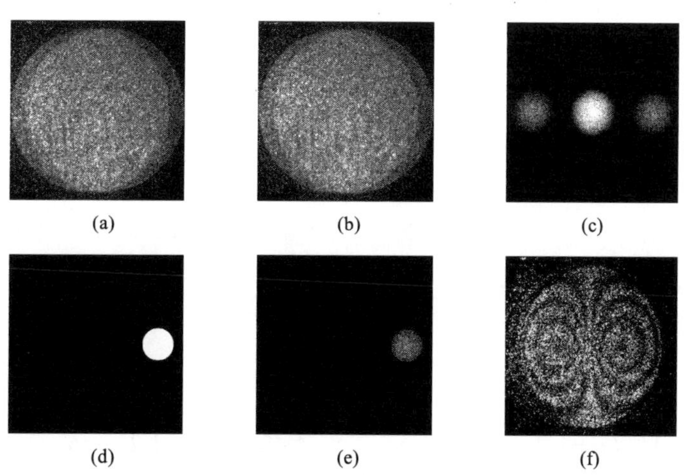

Fig. 6.1 Application of Fourier transform

6.1.2 Discrete Cosine Transform

A discrete cosine transform can express a finite sequence of data points in terms of a sum of cosine functions oscillating at different frequencies. The discrete cosine transform is a transform similar to the discrete Fourier transform, but using only real numbers. The discrete cosine transform is equivalent to the discrete Fourier transforms of roughly twice the length, operating on real data with even symmetry.

The discrete cosine transform represents an image as a sum of sinusoids of varying magnitudes and frequencies. The discrete cosine transform has the property that, for an image, most of the visually significant information about the image is concentrated in just a few coefficients of the discrete cosine transform. For this reason, the discrete cosine transform is often used in image compression, such as compression of JPEG images.

1. Discrete Cosine Transform Principle

1) One-Dimensional Discrete Cosine Transform

In the one-dimensional case, the discrete cosine transform for a signal $f(m)$ of length M is defined as

$$F(p)=C(p)\sum_{m=0}^{M-1}f(m)\cos\left[\frac{\pi(2m+1)p}{2M}\right]\quad(p=0,1,\cdots,M-1)\tag{6.15}$$

where p is the pixel coordinate in the frequency domain, $C(p)=\begin{cases}1/\sqrt{M} & (p=0)\\ \sqrt{2/M} & (p=1,2,\cdots,M-1)\end{cases}$.

The inverse discrete cosine transform is defined as

$$f(m)=C(m)\sum_{p=0}^{M-1}F(p)\cos\left[\frac{\pi(2m+1)p}{2M}\right]\quad(m=0,1,\cdots,M-1)\tag{6.16}$$

where $C(m)=\begin{cases}1/\sqrt{M} & (m=0)\\ \sqrt{2/M} & (m=1,2,\cdots,M-1)\end{cases}$.

2) Two-Dimensional Discrete Cosine Transform

The two-dimensional form of the discrete cosine transform follows immediately from the one-dimensional definition, i.e.,

$$F(p,q)=C(p)C(q)\sum_{m=0}^{M-1}\sum_{n=0}^{N-1}f(m,n)\cos\left[\frac{\pi(2m+1)p}{2M}\right]\cos\left[\frac{\pi(2n+1)q}{2N}\right]\quad(p=0,1,\cdots,M-1;q=0,1,\cdots,N-1)\tag{6.17}$$

where $C(p)=\begin{cases}1/\sqrt{M} & (p=0)\\ \sqrt{2/M} & (p=1,2,\cdots,M-1)\end{cases}$, $C(q)=\begin{cases}1/\sqrt{N} & (q=0)\\ \sqrt{2/N} & (q=1,2,\cdots,N-1)\end{cases}$.

The two-dimensional inverse discrete cosine transform is defined as

$$F(m,n)=C(m)C(n)\sum_{p=0}^{M-1}\sum_{q=0}^{N-1}f(p,q)\cos\left[\frac{\pi(2m+1)p}{2M}\right]\cos\left[\frac{\pi(2n+1)q}{2N}\right] \quad (m=0,1,\cdots,M-1;n=0,1,\cdots,N-1) \tag{6.18}$$

where $C(m)=\begin{cases}1/\sqrt{M} & (m=0)\\ \sqrt{2/M} & (m=1,2,\cdots,M-1)\end{cases}$, $C(n)=\begin{cases}1/\sqrt{N} & (n=0)\\ \sqrt{2/N} & (n=1,2,\cdots,N-1)\end{cases}$.

2. Discrete Cosine Transform Experiment

The discrete cosine transform can also perform the transformation of digital images between the space domain and the frequency domain.

The experimental results of the discrete cosine transform are shown in Fig. 6.2. Fig. 6.2(a) and Fig. 6.2(b) are the digital specklegrams, respectively corresponding to the undeformed and deformed states; Fig. 6.2(c) is the cosine spectrum; Fig. 6.2(d) is the designed ideal band-pass filter; Fig. 6.2(e) is the cosine spectrum after filtering; Fig. 6.2(f) is the contour fringes of out-of-plane displacement derivative.

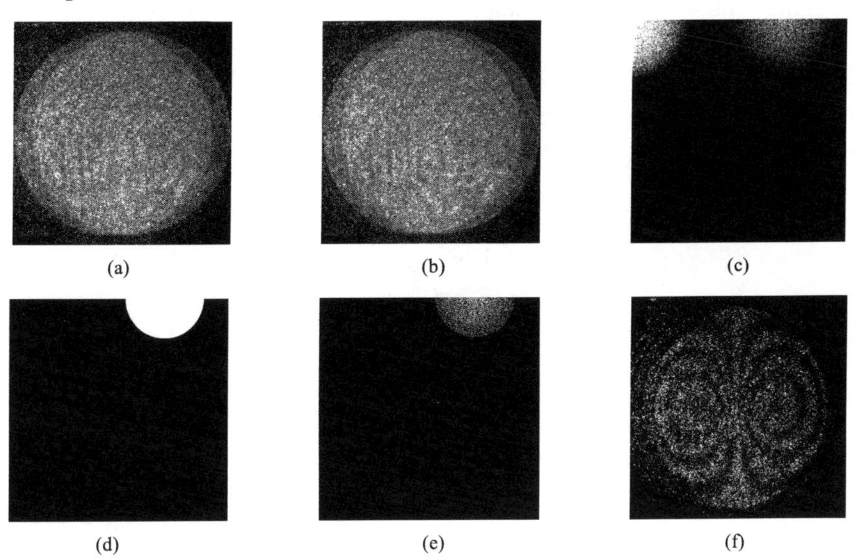

Fig. 6.2 Application of cosine transform

6.2 Low-Pass Filtering

In an interference fringe pattern obtained in optical measurement mechanics, the fringes correspond to a low-frequency component, while the noise, which is randomly distributed over these fringes, corresponds to a high-frequency component. Therefore, a proper low-pass filtering method is usually required to be utilized for denoising.

The low-pass filtering method used for denoising of fringes mainly include averaging, median, and adaptive smooth filtering in the space domain, and ideal, Butterworth, exponential low-pass filtering in the frequency domain.

Three phase-shifted fringe patterns are shown in Fig. 6.3. The corresponding phase-shifting values are 0, $\pi/2$, and π, respectively.

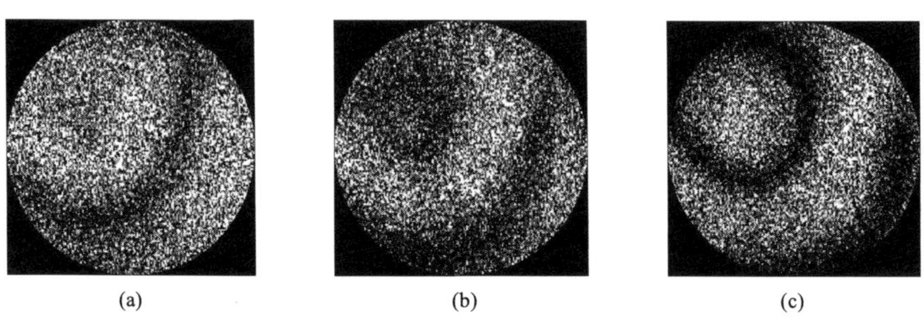

(a) (b) (c)

Fig. 6.3 Fringe patterns

6.2.1 Averaging Smooth Filtering in Space Domain

1. Averaging Smooth Filtering Principle

Averaging smooth filtering is one of the commonly-used smooth filtering techniques used for removing noise from a fringe pattern. Because each pixel gets set to the average value of the pixels in its neighborhood, local variations caused by noise can be reduced. An interference fringe pattern after averaging smooth filtering can be expressed as

$$g(m,n)=\frac{1}{h}\sum_{-p,-q}^{p,q} f(m+i,n+j) \tag{6.19}$$

where $f(m,n)$ and $g(m,n)$ are the gray values of pixel (m,n) before and after filtering, and $h=(2p+1)(2q+1)$ is the size of the filer.

2. Averaging Smooth Filtering Experiment

The filtering results are shown in Fig. 6.4. Fig. 6.4(a), Fig. 6.4(b), and Fig. 6.4(c) are the filtered fringe patterns corresponding to Fig. 6.3(a), Fig. 6.3(b), and Fig. 6.3(c); Fig. 6.4(d) and Fig. 6.4(e) are the $-\pi/2 \sim \pi/2$ and $0 \sim 2\pi$ wrapped phase patterns; Fig. 6.4(f) is the continuous phase pattern.

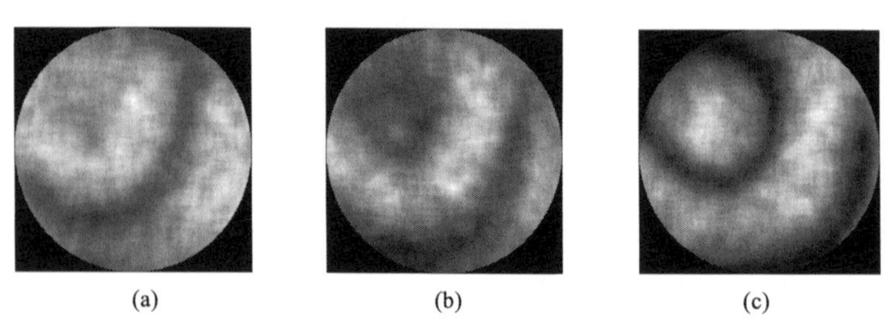

(a) (b) (c)

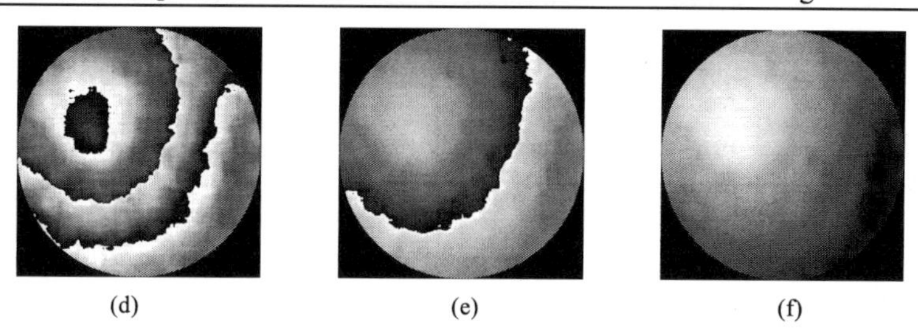

(d) (e) (f)

Fig. 6.4 Averaging smooth filtering results

6.2.2 Median Smooth Filtering in Space Domain

1. Median Smooth Filtering Principle

Median smooth filtering is a nonlinear smooth filtering operation often used in image processing to reduce noise of an interference fringe pattern in the space domain. Each output pixel is determined by the median value in neighborhood around the corresponding pixel in the input image. An interference fringe pattern after median smooth filtering can be expressed as

$$g(m,n)=\mathrm{MF}\{f(m-p,n-q),\cdots,f(m-p,n+q),\cdots,f(m+p,n-q),\cdots,f(m+p,n+q)\} \quad (6.20)$$

where $f(m,n)$ and $g(m,n)$ are the gray values of pixel (m,n) before and after filtering, $(2p+1)(2q+1)$ is the size of the filer, and $\mathrm{MF}\{\cdots\}$ is used to perform the median filtering.

2. Median Smooth Filtering Experiment

The filtering results are shown in Fig. 6.5. Fig. 6.5(a), Fig. 6.5(b), and Fig. 6.5(c) are the filtered fringe patterns corresponding to Fig. 6.3(a), Fig. 6.3(b), and Fig.6.3(c); Fig. 6.5(d) and Fig. 6.5(e) are the $-\pi/2\sim\pi/2$ and $0\sim2\pi$ wrapped phase patterns; Fig. 6.5(f) is the continuous phase pattern.

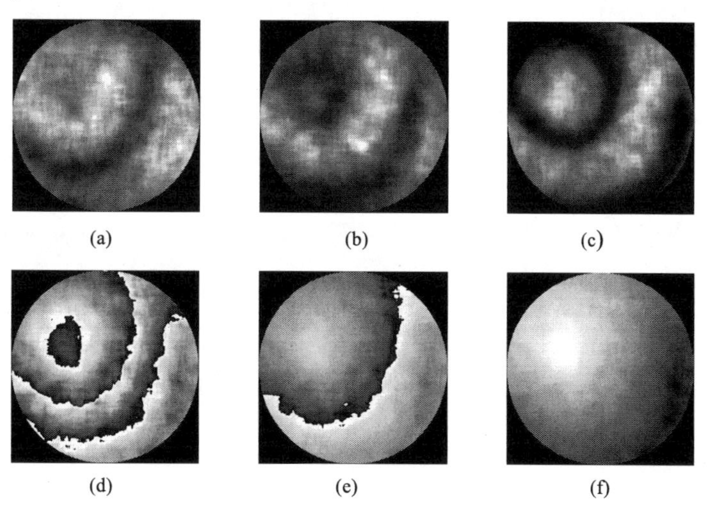

(a) (b) (c)

(d) (e) (f)

Fig. 6.5 Median smooth filtering results

6.2.3　Adaptive Smooth Filtering in Space Domain

1. Adaptive Smooth Filtering Principle

Adaptive smooth filtering uses a pixelwise adaptive Wiener method based on the local image mean and standard deviation estimated from a local neighborhood of each pixel. The adaptive smooth filtering technique can also be used reduce noise of a fringe pattern. The value of an output pixel can be determined by

$$g(m,n)=f(m,n)-\sigma^2\frac{f(m,n)-\frac{1}{h}\sum_{-p,-q}^{p,q}f(m+i,n+j)}{\frac{1}{h}\sum_{-p,-q}^{p,q}[f(m+i,n+j)]^2-\left[\frac{1}{h}\sum_{-p,-q}^{p,q}f(m+i,n+j)\right]^2} \tag{6.21}$$

where $f(m,n)$ and $g(m,n)$ are the gray values of pixel (m,n) before and after filtering, $h=(2p+1)(2q+1)$ is the size of the filer, and σ^2 is the noise variance.

2. Adaptive Smooth Filtering Experiment

The filtering results are shown in Fig. 6.6. Fig. 6.6(a), Fig. 6.6(b), and Fig. 6.6(c) are the filtered fringe patterns corresponding to Fig. 6.3(a), Fig. 6.3(b), and Fig.6.3(c); Fig. 6.6(d) and Fig. 6.6(e) are the $-\pi/2\sim\pi/2$ and $0\sim2\pi$ wrapped phase patterns; Fig. 6.6(f) is the continuous phase pattern.

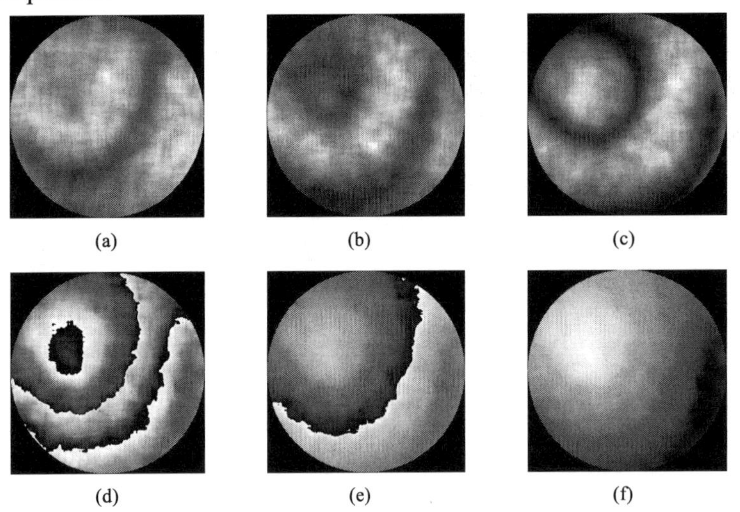

Fig. 6.6　Adaptive smooth filtering results

6.2.4　Ideal Low-Pass Filtering in Frequency Domain

1. Ideal Low-Pass Filtering Principle

Ideal low-pass filtering is the simplest filtering technique in the frequency domain, compared to all the other filtering techniques in the frequency domain, and is often used for denoising of a fringe pattern. An ideal low-pass filter is usually defined as

$$H(p,q)=\begin{cases}1 & (D(p,q)\leqslant D_0)\\ 0 & (D(p,q)>D_0)\end{cases} \tag{6.22}$$

where $H(p,q)$ is the transfer function of the filter, and D_0 is the cut off frequency of the filter.

2. Ideal Low-Pass Filtering Experiment

The filtering results are shown in Fig. 6.7 and Fig.6.8. Fig. 6.7(a), Fig. 6.7(b), and Fig. 6.7(c) are the filtered fringe patterns corresponding to Fig. 6.3(a), Fig. 6.3(b), and Fig. 6.3(c) when subjected to the Fourier transform; Fig. 6.8(a), Fig. 6.8(b), and Fig. 6.8(c) are the filtered fringe patterns corresponding to Fig. 6.3(a), Fig. 6.3(b), and Fig. 6.3(c) when subjected to the cosine transform; Fig. 6.7(d) and Fig. 6.8(d) are the $-\pi/2 \sim \pi/2$ wrapped phase patterns; Fig. 6.7(e) and Fig. 6.8(e) are the $0 \sim 2\pi$ wrapped phase patterns; Fig. 6.7(f) and Fig. 6.8(f) are the continuous phase patterns.

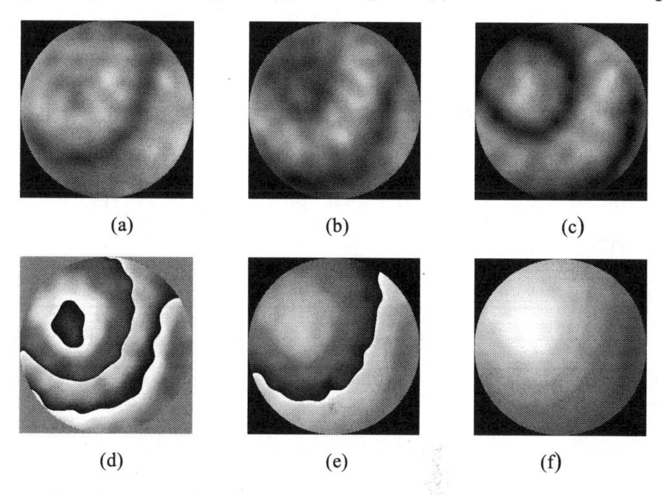

Fig. 6.7 Fourier transform ideal low-pass filtering results

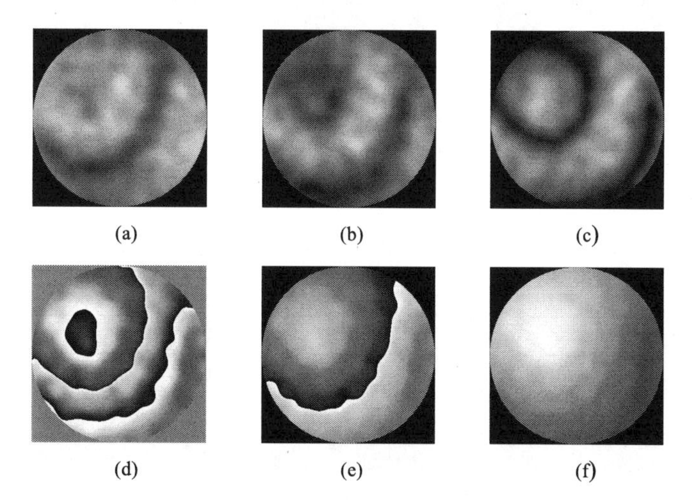

Fig. 6.8 Cosine transform ideal low-pass filtering results

6.2.5 Butterworth Low-Pass Filtering in Frequency Domain

1. Butterworth Low-Pass Filtering Principle

Butterworth low-pass filtering is one of the common-used filtering techniques applied to denoising of a fringe pattern. A Butterworth low-pass filter can be defined as

$$H(p,q)=\frac{1}{1+\left[D(p,q)/D_0\right]^{2n}} \tag{6.23}$$

where D_0 is the cut off frequency of the filter, n is the order of the filter. When $D(p,q)=D_0$, $H(p,q)=0.5$.

2. Butterworth Low-Pass Filtering Experiment

The filtering results are shown in Fig. 6.9 and Fig. 6.10. Fig. 6.9(a), Fig. 6.9(b), and Fig. 6.9(c) are the filtered fringe patterns corresponding to Fig. 6.3(a), Fig. 6.3(b), and Fig. 6.3(c) when subjected to the Fourier transform; Fig. 6.10(a), Fig. 6.10(b), and Fig.6.10(c) are the filtered fringe patterns corresponding to Fig. 6.3(a), Fig. 6.3(b), and Fig. 6.3(c) when subjected to the cosine transform; Fig. 6.9(d) and Fig. 6.10(d) are the $-\pi/2\sim\pi/2$ wrapped phase patterns; Fig. 6.9(e) and Fig. 6.10(e) are the $0\sim2\pi$ wrapped phase patterns; Fig. 6.9(f) and Fig. 6.10(f) are the continuous phase patterns.

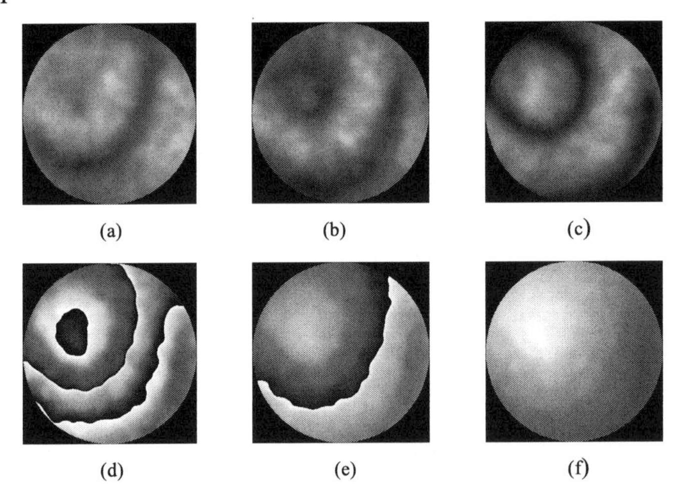

Fig. 6.9　Fourier transform Butterworth low-pass filtering results

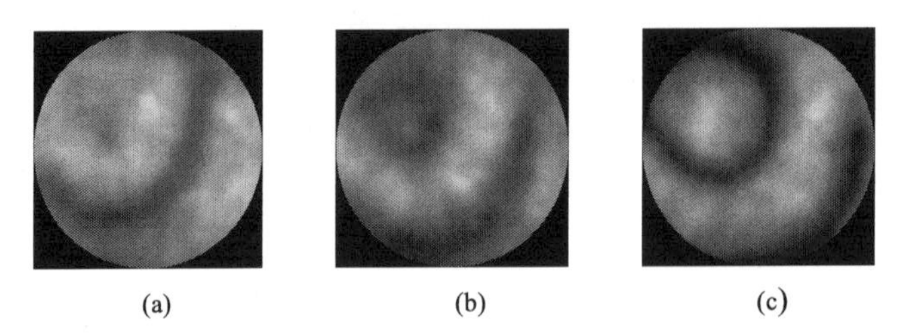

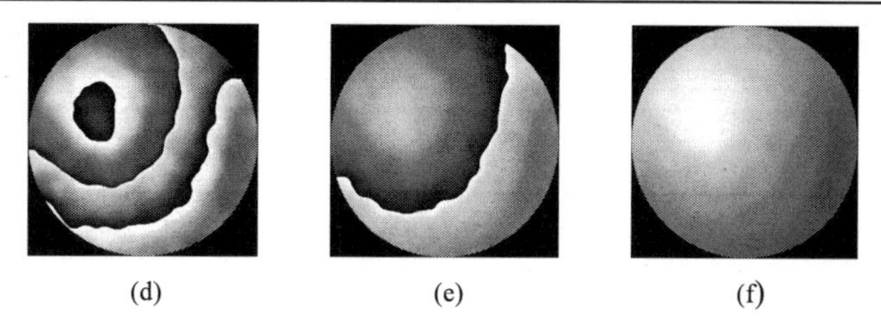

(d) (e) (f)

Fig. 6.10 Cosine transform Butterworth low-pass filtering results

6.2.6 Exponential Low-Pass Filtering in Frequency Domain

1. Exponential Low-Pass Filtering Principle

Because an exponential low-pass filter possesses the function of passing low frequency and of blocking high frequency, it is often applied to denoising of a fringe pattern. An exponential low-pass filter is usually defined as

$$H(p,q)=\exp\left\{-\left[\frac{D(p,q)}{D_0}\right]^n\right\} \tag{6.24}$$

where D_0 is the cut off frequency of the filter, n is the attenuation coefficient of the filter. When $D(p,q)=D_0$, $H(p,q)\approx 1/2.7$.

2. Exponential Low-Pass Filtering Experiment

The filtering results are shown in Fig. 6.11 and Fig. 6.12. Fig. 6.11(a), Fig. 6.11(b), and Fig. 6.11(c) are the filtered fringe patterns corresponding to Fig. 6.3(a), Fig. 6.3(b), and Fig. 6.3(c) when subjected to the Fourier transform; Fig. 6.12(a), Fig. 6.12(b), and Fig.6.12(c) are the filtered fringe patterns corresponding to Fig. 6.3(a), Fig. 6.3(b), and Fig. 6.3(c) when subjected to the cosine transform; Fig. 6.11(d) and Fig. 6.12(d) are the $-\pi/2\sim\pi/2$ wrapped phase patterns; Fig. 6.11(e) and Fig. 6.12(e) are the $0\sim 2\pi$ wrapped phase patterns; Fig. 6.11(f) and Fig. 6.12(f) are the continuous phase patterns.

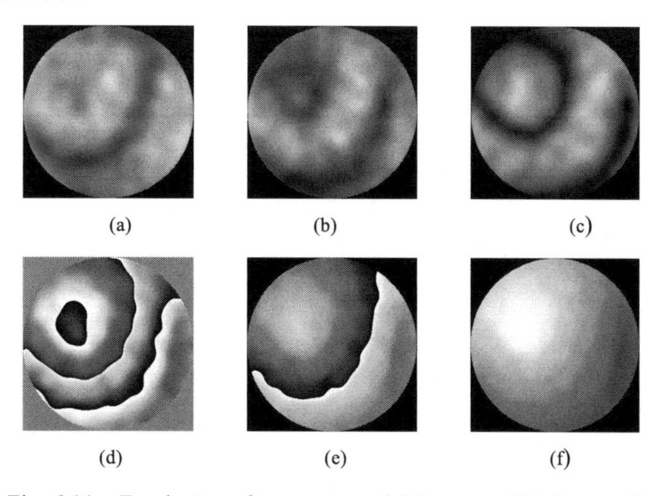

(a) (b) (c)

(d) (e) (f)

Fig. 6.11 Fourier transform exponential low-pass filtering results

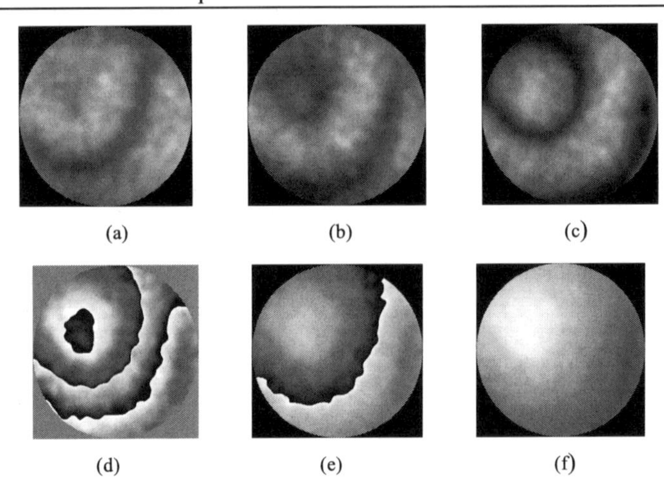

Fig. 6.12　Cosine transform exponential low-pass filtering results

Chapter 7 Digital Holography and Digital Holographic Interferometry

Holography is to record interference fringes formed by interference of the object and the reference waves. Because the spatial frequency of these interference fringes is usually very high, the recording medium needs to have high resolution. Since holography was proposed, photographic plates with high resolution have been used as recording medium. However, since a photographic plate has low sensitivity and needs long exposure time, the stability of recording system is highly required. In addition, after this photographic plate records holograms, it needs to be subjected to a wet processing, such as development, fixation, etc. In order to overcome these shortcomings in holography, digital holography was proposed. In digital holography, a photosensitive electronic device (such as CCD or CMOS) is used to record holograms, instead of a traditional recording medium. Using a photosensitive electronic device as recording medium avoids wet processing, such as development, fixation, etc.; thus the recording process is greatly simplified. In addition, the digital method can be used to simulate light wave diffraction to reconstruct the object wave; thus an optical reconstruction setup can be left out from digital holography.

Although digital holography has many advantages, it also has some shortcomings. Compared with the traditional recording medium (such as photographic plate), the spatial resolution of the photosensitive electronic device is still low and the size of the photosensitive surface is relatively small; thus the resolution of reconstructed images in digital holographic is not high at the present time. However, with rapid development of computer science and photosensitive electronic devices, these problems will be gradually solved, and digital holography will be more greatly developed and more widely used.

7.1 Digital Holography

The basic theory and experimental techniques of optical holography are also applicable to digital holography. However, since the photosensitive electronic device used for recording digital holograms has relatively small photosensitive surface and low spatial resolution at present, digital holography can only record and reconstruct small objects within a limited distance. Digital holography needs to satisfy all the requirements in optical holography, and it is also needed to meet the sampling theorem in recording process.

The commonly-used photosensitive electronic device in digital holographic recording is a CCD. As recording medium, the CCD has a high photosensitivity and a wide response range of wavelength, so it has a great advantage when used for holographic recording. In addition, because the CCD is relatively cheap, it has been widely used in digital holography.

7.1.1 Digital Holographic Recording

A digital holographic recording system is basically the same as a traditional holographic recording system. The main difference between these two systems is that a CCD is used to replace a photographic plate as recording medium. The off-axis digital holographic recording system is shown in Fig. 7.1. The light wave from a laser is split into two beams by a beamsplitter. One of the beams, called the object light, is used to illuminate the object after it is reflected by mirror and expanded by beam expander, and then the CCD target is illuminated by the beam scattered from the surface of object. The other beam, called the reference wave, is used to directly illuminate the CCD target after it is reflected by mirror and expanded by beam expander. The object and reference light beams are superposed coherently on the CCD target to form a Fresnel hologram.

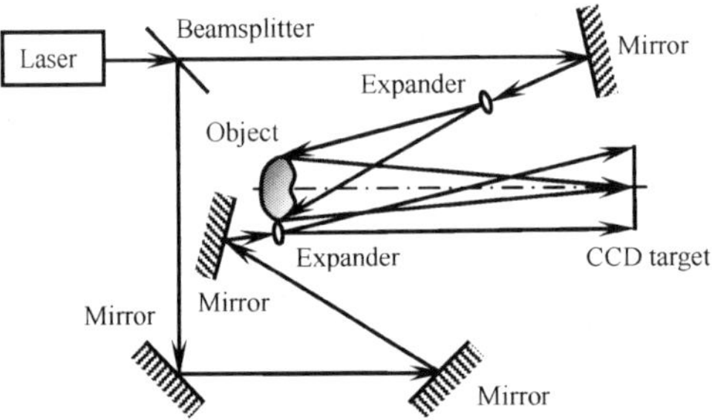

Fig. 7.1　Off-axis recording system

Assuming that the complex amplitude of light wave on the object surface is denoted by $O(x_o, y_o)$, then in the Fresnel diffraction region, the complex amplitude of object light wave on the CCD target can be expressed as

$$O(x,y)=\frac{\exp(\mathrm{i}kz_o)}{\mathrm{i}\lambda z_o}\exp\left[\mathrm{i}\frac{\pi}{\lambda z_o}(x^2+y^2)\right]\int_{-\infty}^{\infty}\int_{-\infty}^{\infty}O(x_o,y_o) \cdot\exp\left[\mathrm{i}\frac{\pi}{\lambda z_o}(x_o^2+y_o^2)\right]\exp\left[-\mathrm{i}\frac{2\pi}{\lambda z_o}(xx_o+yy_o)\right]\mathrm{d}x_o\mathrm{d}y_o \tag{7.1}$$

where $\mathrm{i}=\sqrt{-1}$ is the imaginary unit, λ is the wavelength, $k=2\pi/\lambda$ is the wave number, and z_o is the distance from the object surface to the CCD target.

Assuming that the complex amplitude of the reference light wave on the CCD target is given by $R(x,y)$, then the intensity distribution on the CCD target, generated by the interference of the object light wave and the reference wave, can be expressed as

$$I(x,y)=|O(x,y)|^2+|R(x,y)|^2+O(x,y)R^*(x,y)+O^*(x,y)R(x,y) \tag{7.2}$$

where * represents complex conjugate, $|O(x,y)|^2$ and $|R(x,y)|^2$ are respectively the intensity distribution of the object light and the reference wave, $O(x,y)R^*(x,y)$ and

$O^*(x,y)R(x,y)$ are both the interference terms, containing the amplitude and phase information of the object light wave, respectively.

The hologram recorded by a CCD camera is called the digital hologram. Assuming that the effective pixels of the CCD are $M \times N$ and that the center distances of the adjacent pixels are Δx and Δy in the x and y directions, the discrete intensity distribution of the digital hologram recorded by the CCD is

$$I(m,n) = I(x,y)S(m,n) = \sum_{m=0}^{M-1}\sum_{n=0}^{N-1} I(x,y)\delta(x - m\Delta x, y - n\Delta y) \tag{7.3}$$

where

$$S(m,n) = \sum_{m=0}^{M-1}\sum_{n=0}^{N-1} \delta(x - m\Delta x, y - n\Delta y) \tag{7.4}$$

is the sampling function.

7.1.2 Digital Holographic Reconstruction

In digital holography, the numerical method is used to simulate the diffraction of light wave and the reconstruction of object light. Assuming that the complex amplitude of the reconstruction wave is denoted by $C(x,y)$, the complex amplitude of the light wave through the hologram can be expressed as

$$A(m,n) = C(x,y)I(m,n) = \sum_{m=0}^{M-1}\sum_{n=0}^{N-1} C(x,y)I(x,y)\delta(x - m\Delta x, y - n\Delta y) \tag{7.5}$$

In the Fresnel diffraction region, the complex amplitude distribution of the diffraction light, having the distance z_{r} from the hologram, can be written by

$$\begin{aligned} A(p,q) = {} & \frac{\exp(\mathrm{i}kz_{\mathrm{r}})}{\mathrm{i}\lambda z_{\mathrm{r}}} \exp\left[\mathrm{i}\frac{\pi}{\lambda z_{\mathrm{r}}}(p^2\Delta x_{\mathrm{r}}^2 + q^2\Delta y_{\mathrm{r}}^2)\right] \\ & \times \sum_{m=0}^{M-1}\sum_{n=0}^{N-1} A(m,n)\exp\left[\mathrm{i}\frac{\pi}{\lambda z_{\mathrm{r}}}(m^2\Delta x^2 + n^2\Delta y^2)\right]\exp\left[-\mathrm{i}2\pi\left(\frac{mp}{M} + \frac{nq}{N}\right)\right] \\ & (p = 0,1,\cdots,M-1; q = 0,1,\cdots,N-1) \end{aligned} \tag{7.6}$$

where Δx_{r} and Δy_{r} are the center distances between adjacent pixels on the observation plane having the distance z_{r} from the hologram in the x and y directions, and can be given by

$$\Delta x_{\mathrm{r}} = \frac{\lambda z_{\mathrm{r}}}{M\Delta x}, \quad \Delta y_{\mathrm{r}} = \frac{\lambda z_{\mathrm{r}}}{N\Delta y} \tag{7.7}$$

Substituting Eq. (7.7) into Eq. (7.6), we obtain

$$\begin{aligned} A(p,q) = {} & \frac{\exp(\mathrm{i}kz_{\mathrm{r}})}{\mathrm{i}\lambda z_{\mathrm{r}}} \exp\left[\mathrm{i}\pi\lambda z_{\mathrm{r}}\left(\frac{p^2}{M^2\Delta x^2} + \frac{q^2}{N^2\Delta y^2}\right)\right] \\ & \times \sum_{m=0}^{M-1}\sum_{n=0}^{N-1} A(m,n)\exp\left[\mathrm{i}\frac{\pi}{\lambda z_{\mathrm{r}}}(m^2\Delta x^2 + n^2\Delta y^2)\right]\exp\left[-\mathrm{i}2\pi\left(\frac{mp}{M} + \frac{nq}{N}\right)\right] \\ & (p = 0,1,\cdots,M-1; q = 0,1,\cdots,N-1) \end{aligned} \tag{7.8}$$

Therefore, the intensity and phase distributions are

$$I(p,q) = |A(p,q)|^2 \quad (p = 0,1,\cdots,M-1; q = 0,1,\cdots,N-1) \tag{7.9}$$

and

$$\varphi(p,q) = \arctan\frac{\text{Im}\{A(p,q)\}}{\text{Re}\{A(p,q)\}} \quad (p = 0,1,\cdots,M-1; q = 0,1,\cdots,N-1) \tag{7.10}$$

where Re{·} and Im{·} are respectively the real part and the imaginary part. For rough surfaces, the phase $\varphi(p,q)$ is randomly changed, and the intensity $I(p,q)$ in digital holography is of interest.

In digital holography, the in-line digital holographic recording system shown in Fig. 7.2 is also an often-used system.

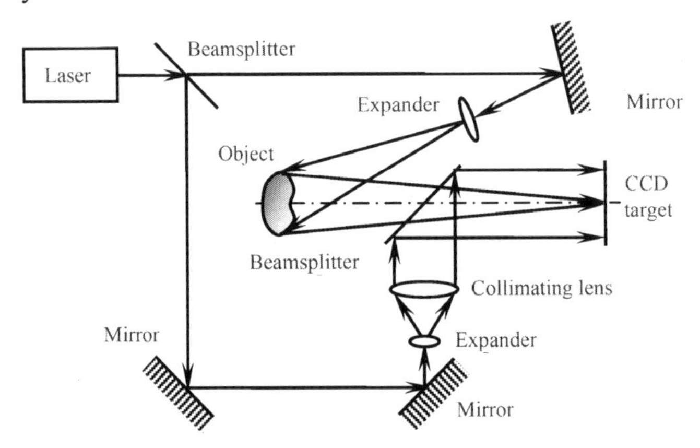

Fig. 7.2 In-line recording system

7.1.3 Digital Holographic Experiments

Expt. 1 The digital hologram is shown in Fig. 7.3(a), and the reconstructed image is shown in Fig. 7.3(b).

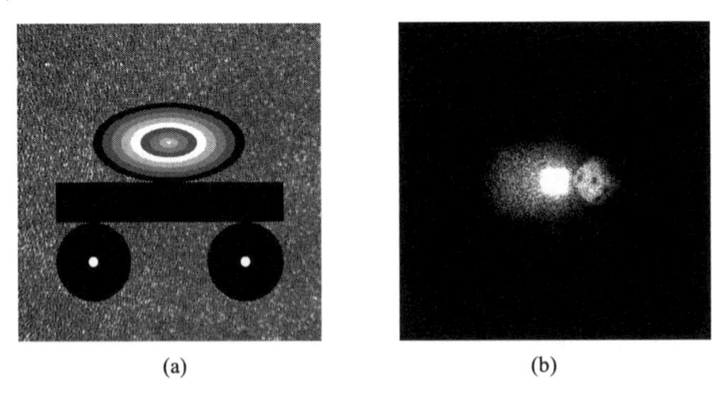

Fig. 7.3 Reconstruction result obtained in Expt. 1

Expt. 2 The digital hologram is shown in Fig. 7.4(a), and the reconstructed image is shown in Fig. 7.4(b).

(a)

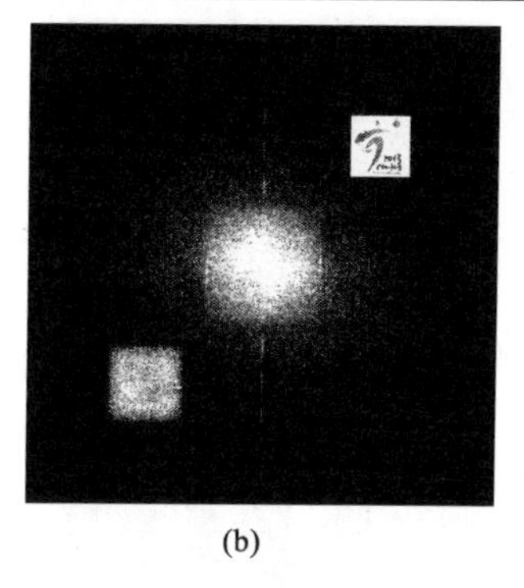

(b)

Fig. 7.4 Reconstruction result obtained in Expt. 2

7.2 Digital Holographic Interferometry

1. Principle of Digital Holographic Interferometry

Digital holographic interferometry, similar to conventional holographic interferometry, can also be used to measure the deformation of object. Assuming that the complex amplitude distributions before and after deformation are respectively represented by $A_1(p,q)$ and $A_2(p,q)$, the phase distributions before and after deformation can be expressed as

$$\varphi_1(p,q)=\arctan\frac{\mathrm{Im}\{A_1(p,q)\}}{\mathrm{Re}\{A_1(p,q)\}},\quad \varphi_2(p,q)=\arctan\frac{\mathrm{Im}\{A_2(p,q)\}}{\mathrm{Re}\{A_2(p,q)\}}\\ (p=0,1,\cdots,M-1;q=0,1,\cdots,N-1) \tag{7.11}$$

The phase distributions of the object before and after deformation are random variables, but their difference will no longer be a random variable, represents the phase change caused by the load applied to the object, and is related to the object deformation alone. Using Eq. (7.11), the phase change caused by the deformation of object can be expressed as

$$\delta(p,q)=\varphi_2(p,q)-\varphi_1(p,q)=\arctan\frac{\mathrm{Im}\{A_2(p,q)\}}{\mathrm{Re}\{A_2(p,q)\}}-\arctan\frac{\mathrm{Im}\{A_1(p,q)\}}{\mathrm{Re}\{A_1(p,q)\}}\\ (p=0,1,\cdots,M-1;q=0,1,\cdots,N-1) \tag{7.12}$$

where $\delta(p,q)$ is a wrapped phase. A continuous phase distribution can be obtained by phase unwrapping.

2. Experiment of Digital Holographic Interferometry

Phase distributions obtained in digital holographic interferometry are shown in Fig. 7.5. Fig. 7.5(a) and Fig. 7.5(b) are phase distributions before and after deformation; Fig. 7.5(c) is a wrapped phase distribution related to the object deformation, which is obtained by subtracting Fig. 7.5(a) from Fig. 7.5(b); Fig. 7.5(d) is a continuous phase distribution obtained by unwrapping the wrapped phase of Fig. 7.5(c).

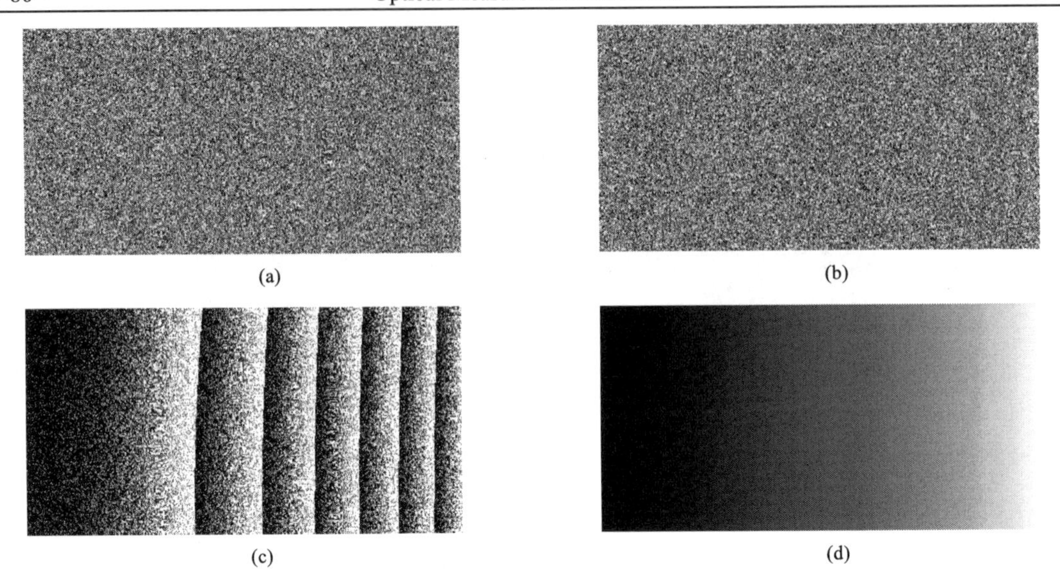

(a) (b) (c) (d)

Fig. 7.5　Phase distributions

Chapter 8 Digital Speckle Interferometry and Digital Speckle Shearing Interferometry

The basic principle of digital speckle (shearing) interferometry is the same as that of traditional speckle (shearing) interferometry. The traditional method uses photographic plates to record specklegrams, hence wet processing such as development and fixation is needed; in addition, in the traditional method two series of exposure are added, so filtering is needed so as to obtain interference fringes. The digital method uses CCD cameras to record digital specklegrams, so wet processing including development and fixation is no longer needed; in addition, digital specklegrams can be stored additively, or can also be stored separately, so besides the additive method, the digital method can also use the subtraction method.

In digital speckle (shearing) interferometry, the background intensity can be removed by subtracting two exposure recordings before and after deformation, so the subtraction method is usually adopted in digital speckle (shearing) interferometry. The subtraction method can obtain directly interference fringes without requiring filtering. Two series of exposure before and after deformation are processed independently, therefore phase-shifting interferometry can be easily applied to digital speckle (shearing) interferometry.

8.1 Digital Speckle Interferometry

8.1.1 In-Plane Displacement Measurement

The digital speckle interferometric system for measuring in-plane displacement is shown in Fig. 8.1. Two beams of collimated waves, having the same included angle θ measured from the normal of the surface, are used to illuminate symmetrically the object plane. The scattered light waves are superposed coherently on the CCD target.

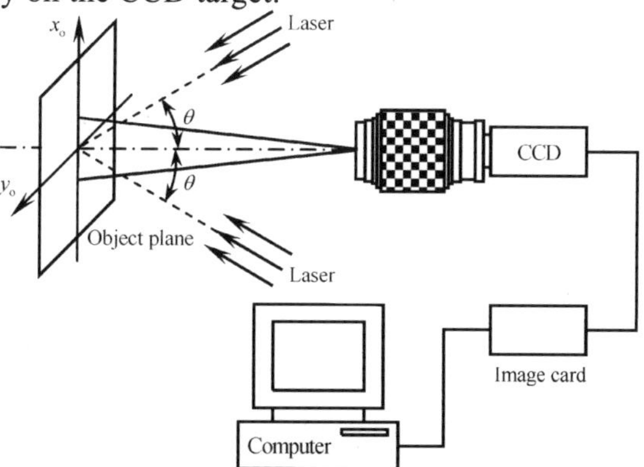

Fig. 8.1 In-plane displacement measuring system

1. Fringe Formation

The intensity distribution recorded by the CCD before the object deformation can be expressed as

$$I_1 = I_{o1} + I_{o2} + 2\sqrt{I_{o1}I_{o2}}\cos\varphi \tag{8.1}$$

where I_{o1} and I_{o2} are respectively the intensity distributions of the two incident waves on the CCD target, φ is the phase difference between the two incident waves.

The intensity distribution recorded by the CCD after the deformation of object can be given by

$$I_2 = I_{o1} + I_{o2} + 2\sqrt{I_{o1}I_{o2}}\cos(\varphi + \delta) \tag{8.2}$$

where δ is the relative phase change for the two incident waves, which is caused by the deformation of object, and can be expressed as

$$\delta = 2ku_o \sin\theta \tag{8.3}$$

where k is the wave number, u_o is the in-plane displacement component of the object surface along the x direction.

The squared difference of the intensity distributions recorded before and after deformation can be written as

$$E = (I_2 - I_1)^2 = 8I_{o1}I_{o2}\sin^2\left(\varphi + \frac{1}{2}\delta\right)(1 - \cos\delta) \tag{8.4}$$

where the sine term is a high frequency component, which corresponds to the speckle noise, whereas the cosine term is a low frequency component, which corresponds to the deformation of object. Therefore, when the condition

$$\delta = 2n\pi \quad (n = 0, \pm 1, \pm 2, \cdots) \tag{8.5}$$

is met, the brightness of fringes will reach minimum; i.e., dark fringes will be formed when

$$u_o = \frac{n\pi}{k\sin\theta} = \frac{n\lambda}{2\sin\theta} \quad (n = 0, \pm 1, \pm 2, \cdots) \tag{8.6}$$

and when the condition

$$\delta = (2n+1)\pi \quad (n = 0, \pm 1, \pm 2, \cdots) \tag{8.7}$$

is met, the brightness of fringes will reach maximum; i.e., bright fringes will be formed when

$$u_o = \frac{(2n+1)\pi}{2k\sin\theta} = \frac{(2n+1)\lambda}{4\sin\theta} \quad (n = 0, \pm 1, \pm 2, \cdots) \tag{8.8}$$

2. Phase Analysis

A digital specklegram is first recorded before the deformation of object, and then four digital specklegrams having phase-shifting values 0, $\pi/2$, π and $3\pi/2$ are recorded, respectively, after the deformation of object. Through digital subtraction of the specklegram before deformation from the specklegrams after deformation, the squared differences can be written by

$$
\begin{aligned}
E_1 &= 8I_{o1}I_{o2}\sin^2\left(\varphi+\frac{1}{2}\delta\right)(1-\cos\delta) \\
E_2 &= 8I_{o1}I_{o2}\sin^2\left(\varphi+\frac{1}{2}\delta+\frac{1}{4}\pi\right)(1+\sin\delta) \\
E_3 &= 8I_{o1}I_{o2}\sin^2\left(\varphi+\frac{1}{2}\delta+\frac{1}{2}\pi\right)(1+\cos\delta) \\
E_4 &= 8I_{o1}I_{o2}\sin^2\left(\varphi+\frac{1}{2}\delta+\frac{3}{4}\pi\right)(1-\sin\delta)
\end{aligned}
\tag{8.9}
$$

The sine terms in Eq. (8.9) correspond to the high frequency noise, and can be removed by low-pass filtering, thus Eq. (8.9) can be rewritten by

$$
\begin{aligned}
<E_1> &= 4<I_{o1}><I_{o2}>(1-\cos\delta) \\
<E_2> &= 4<I_{o1}><I_{o2}>(1+\sin\delta) \\
<E_3> &= 4<I_{o1}><I_{o2}>(1+\cos\delta) \\
<E_4> &= 4<I_{o1}><I_{o2}>(1-\sin\delta)
\end{aligned}
\tag{8.10}
$$

where $<\cdot>$ denotes the ensemble average. By solving the above equations, the phase distribution of the object deformation can be expressed as

$$
\delta = \arctan\frac{S}{C} = \arctan\frac{<E_2>-<E_4>}{<E_3>-<E_4>}
\tag{8.11}
$$

where δ is the wrapped phase in the $-\pi/2 \sim \pi/2$ range, $S=<E_2>-<E_4>$, $C=<E_3>-<E_1>$. The above wrapped phase can be extended to the $0\sim 2\pi$ range by the following transformation:

$$
\delta = \begin{cases}
\delta & (S \geqslant 0,\ C>0) \\
\frac{1}{2}\pi & (S>0,\ C=0) \\
\delta+\pi & (C<0) \\
\frac{3}{2}\pi & (S<0,\ C=0) \\
\delta+2\pi & (S<0,\ C>0)
\end{cases}
\tag{8.12}
$$

The phase distribution obtained from Eq. (8.12) is still a wrapped phase, thus it is necessary to unwrap the above wrapped phase so as to obtain a continuous phase distribution. If the phase difference between adjacent pixels reaches or exceeds π, then the phase is increased or decreased by $2n\pi$ so as to eliminate the discontinuity of phase. Thus the unwrapped phase can be expressed as

$$
\delta_u = \delta + 2n\pi \quad (n=\pm1,\pm2,\pm3,\cdots)
\tag{8.13}
$$

where δ_u denotes the continuous phase distribution. From the continuous phase distribution obtained, the distribution of in-plane displacement can be expressed as

$$
u_o = \frac{\delta_u}{2k\sin\theta} = \frac{\lambda\delta_u}{4\pi\sin\theta}
\tag{8.14}
$$

3. Experimental Verification

Four patterns of contour fringes with phase-shifting values 0, π/2, π and 3π/2, obtained in in-plane displacement measurement, are as shown in Fig. 8.2.

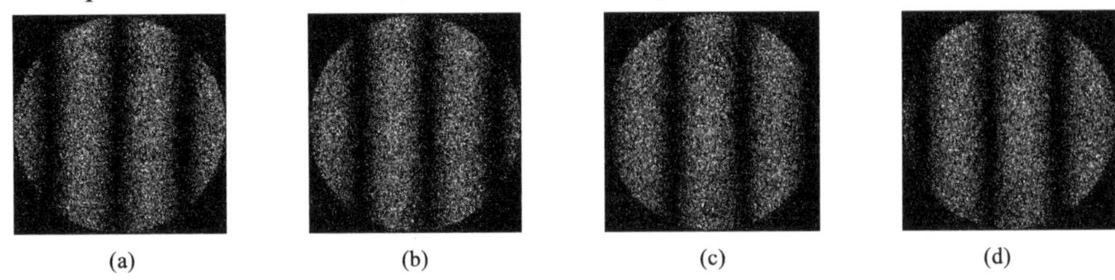

Fig. 8.2　In-plane displacement contour fringes

Two phase maps for in-plane displacement are shown in Fig. 8.3. Fig. 8.3(a) is the phase map wrapped in the $0\sim2\pi$ range; Fig. 8.3(b) is the continuous phase map.

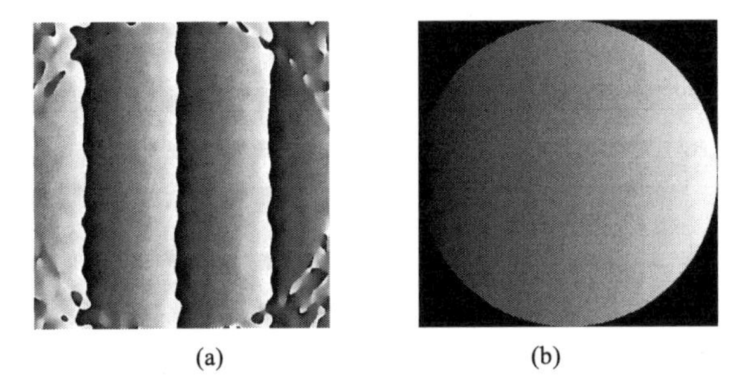

Fig. 8.3　Phase maps for in-plane displacement

8.1.2 Out-of-Plane Displacement Measurement

The out-of-plane displacement measurement system is shown in Fig. 8.4.

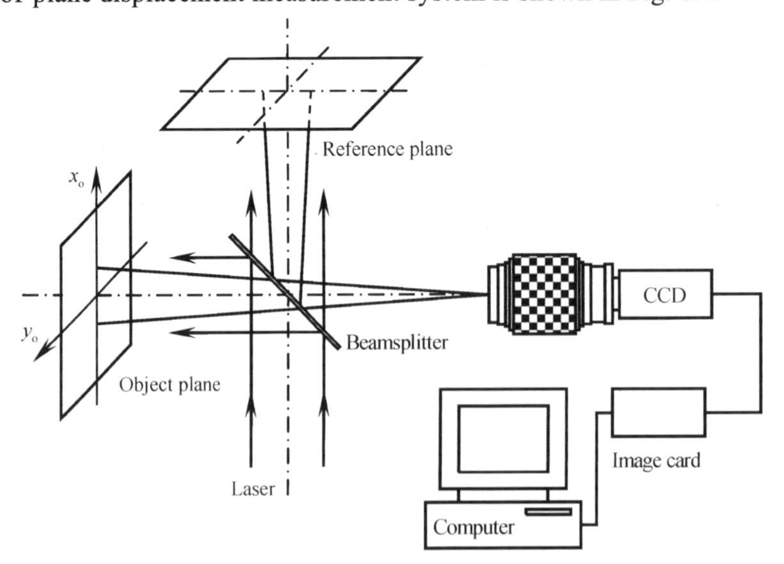

Fig. 8.4　Out-of-plane displacement measuring system

1. Fringe Formation

The intensity distribution recorded before deformation can be expressed as

$$I_1 = I_o + I_r + 2\sqrt{I_o I_r}\cos\varphi \tag{8.15}$$

where I_o and I_r are respectively the intensity distributions of the object and reference lights, and φ is the phase difference between the object and reference waves.

Similarly, the intensity distribution recorded after deformation can be given by

$$I_2 = I_o + I_r + 2\sqrt{I_o I_r}\cos(\varphi + \delta) \tag{8.16}$$

where

$$\delta = 2kw_o \tag{8.17}$$

where w_o is the out-of-plane displacement component.

Using the subtraction method, the squared difference of the two digital specklegrams can be written as

$$E = (I_2 - I_1)^2 = 8I_o I_r \sin^2\left(\varphi + \frac{1}{2}\delta\right)(1 - \cos\delta) \tag{8.18}$$

Therefore, when the condition

$$\delta = 2n\pi \quad (n = 0, \pm 1, \pm 2, \cdots) \tag{8.19}$$

is satisfied, the brightness of fringes will be minimum, i.e., dark fringes will be generated when

$$w_o = \frac{n\pi}{k} = \frac{n\lambda}{2} \quad (n = 0, \pm 1, \pm 2, \cdots) \tag{8.20}$$

And when the condition

$$\delta = (2n+1)\pi \quad (n = 0, \pm 1, \pm 2, \cdots) \tag{8.21}$$

is satisfied, the brightness of fringes will be maximum, i.e., bright fringes will be formed when

$$w_o = \frac{(2n+1)\pi}{2k} = \frac{(2n+1)\lambda}{4} \quad (n = 0, \pm 1, \pm 2, \cdots) \tag{8.22}$$

2. Phase Analysis

First record a digital specklegram before the deformation of object; then record respectively four digital specklegrams having phase-shifting values 0, $\pi/2$, π and $3\pi/2$ after the deformation of object. Through digital subtraction of the specklegram before deformation from the digital specklegrams after deformation, four contour fringe patterns with phase-shifting values 0, $\pi/2$, π and $3\pi/2$ can be obtained. When these four contour fringe patterns is subjected to low-pass filtering, the ensemble average of these patterns can be expressed as

$$\begin{aligned} &\langle E_1 \rangle = 4\langle I_o \rangle\langle I_r \rangle(1 - \cos\delta) \\ &\langle E_2 \rangle = 4\langle I_o \rangle\langle I_r \rangle(1 + \sin\delta) \\ &\langle E_3 \rangle = 4\langle I_o \rangle\langle I_r \rangle(1 + \cos\delta) \\ &\langle E_4 \rangle = 4\langle I_o \rangle\langle I_r \rangle(1 - \sin\delta) \end{aligned} \tag{8.23}$$

By solving the above equations, the phase distribution related to the object deformation can be expressed as

$$\delta = \arctan \frac{<E_2> - <E_4>}{<E_3> - <E_1>} \tag{8.24}$$

where δ is the wrapped phase in the $-\pi/2 \sim \pi/2$ range. A continuous phase distribution can be obtained by phase unwrapping. From the continuous phase distribution, the distribution of out-of-plane displacement can be expressed as

$$w_o = \frac{\delta_u}{2k} = \frac{\lambda \delta_u}{4\pi} \tag{8.25}$$

3. Experimental Certification

Four contour fringe patterns with phase-shifting values 0, $\pi/2$, π and $3\pi/2$, which are obtained in out-of-plane displacement measurement, are shown in Fig. 8.5.

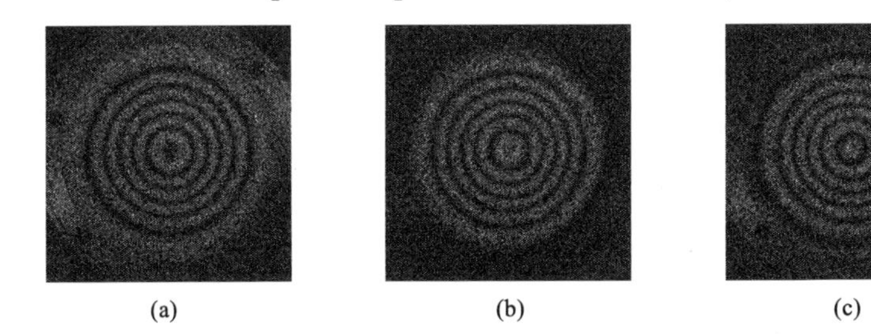

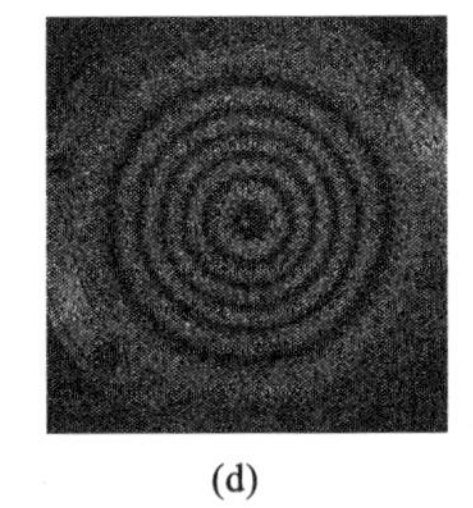

(a)　(b)　(c)　(d)

Fig. 8.5　Out-of-plane displacement contour fringes

Two phase maps for out-of-plane displacement are shown in Fig. 8.6. Fig. 8.6(a) is the wrapped phase map of $0 \sim 2\pi$ range; Fig. 8.6(b) is the continuous phase map.

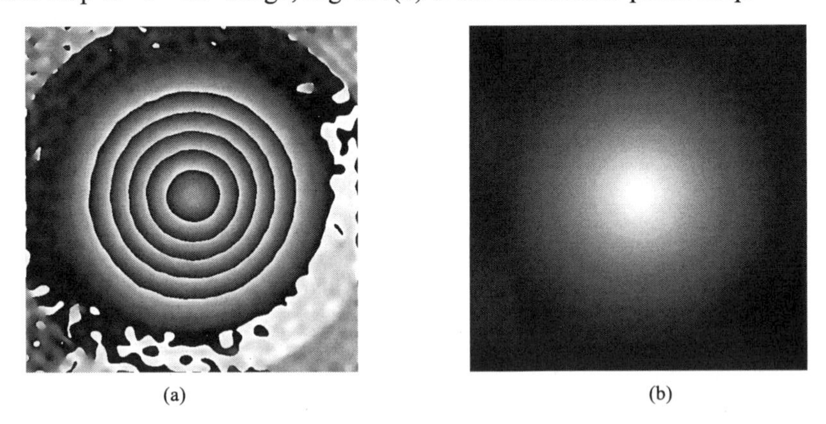

(a)　(b)

Fig. 8.6　Phase maps for out-of-plane displacement

8.2　Digital Speckle Shearing Interferometry

The digital speckle shearing system is shown in Fig. 8.7. The object is illuminated by a collimated laser beam, and the light scattered from the object surface is focused on the CCD target. By tilting one of the flat mirrors, two speckle fields will be superposed coherently to form a resultant speckle field.

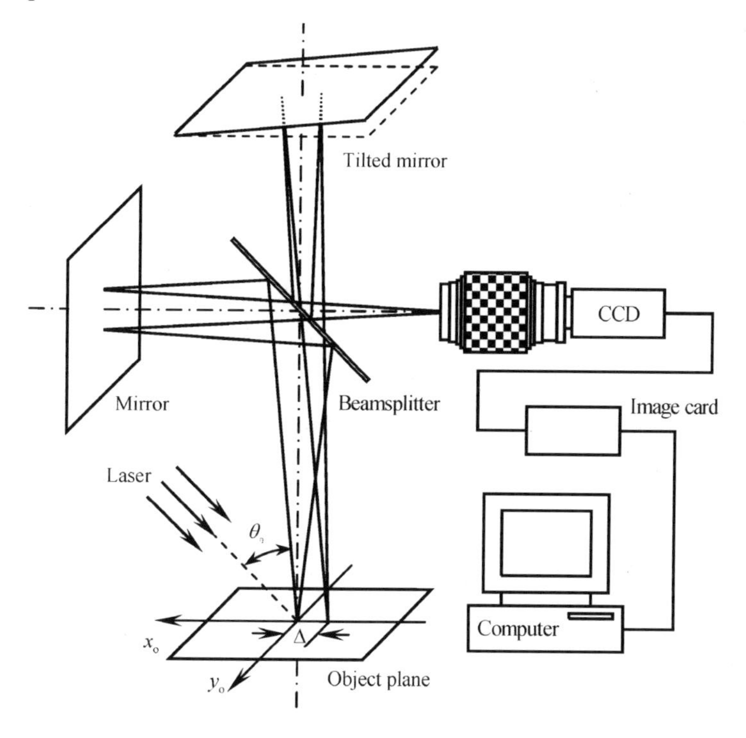

Fig. 8.7 Digital speckle shearing system

1. Fringe Formation

The intensity distribution before deformation recorded by the CCD can be expressed as

$$I_1 = I_{o1} + I_{o2} + 2\sqrt{I_{o1}I_{o2}}\cos\varphi \tag{8.26}$$

where I_{o1} and I_{o2} are respectively the intensity distributions of the two sheared speckle fields, φ is the phase difference between the two speckle fields.

Similarly, the intensity distribution after deformation recorded by the CCD can be given by

$$I_2 = I_{o1} + I_{o2} + 2\sqrt{I_{o1}I_{o2}}\cos(\varphi + \delta) \tag{8.27}$$

When the incident laser beam is perpendicular to the object plane, then we have

$$\delta = 2k\frac{\partial w_o}{\partial x}\varDelta \tag{8.28}$$

where $\varDelta$ is the shearing value of the two speckle fields on the object plane, and $\partial w_o/\partial x$ is the derivative of out-of-plane displacement with respect to x (i.e., slope).

The squared difference of the intensity distributions recorded respectively before and after deformation can be expressed as

$$E=(I_2-I_1)^2=8I_{o1}I_{o2}\sin^2(\varphi+\tfrac{1}{2}\delta)(1-\cos\delta) \tag{8.29}$$

Obviously, dark fringes will be produced when $\delta=2n\pi$, i.e.,

$$\frac{\partial w_o}{\partial x}=\frac{n\pi}{k\Delta}=\frac{n\lambda}{2\Delta}\quad (n=0,\pm1,\pm2,\cdots) \tag{8.30}$$

and bright fringes will be formed when $\delta=(2n+1)\pi$, or

$$\frac{\partial w_o}{\partial x}=\frac{(2n+1)\pi}{2k\Delta}=\frac{(2n+1)\lambda}{4\Delta}\quad (n=0,\pm1,\pm2,\cdots) \tag{8.31}$$

2. Phase Analysis

Assume that one specklegram is recorded before deformation and that four specklegrams, having phase-shifting values -3α, $-\alpha$, α and 3α (α is a constant), are recorded after deformation. When the specklegram before deformation is subtracted digitally from the specklegrams after deformation, we can obtain four contour fringe patterns of slope with phase-shifting values -3α, $-\alpha$, α and 3α. These patterns, when subjected to low-pass filtering, can be expressed as

$$\begin{aligned}
<E_1>&=4<I_{o1}><I_{o2}>[1-\cos(\delta-3\alpha)]\\
<E_2>&=4<I_{o1}><I_{o2}>[1-\cos(\delta-\alpha)]\\
<E_3>&=4<I_{o1}><I_{o2}>[1-\cos(\delta+\alpha)]\\
<E_4>&=4<I_{o1}><I_{o2}>[1-\cos(\delta+3\alpha)]
\end{aligned} \tag{8.32}$$

By solving the above equations, the phase distribution related to the object deformation can be expressed as

$$\delta=\arctan\left[\tan\beta\frac{(<E_2>-<E_3>)+(<E_1>-<E_4>)}{(<E_2>+<E_3>)-(<E_1>+<E_4>)}\right] \tag{8.33}$$

where β can be given by

$$\beta=\arctan\sqrt{\frac{3(<E_2>-<E_3>)-(<E_1>-<E_4>)}{(<E_2>-<E_3>)+(<E_1>-<E_4>)}} \tag{8.34}$$

The phase obtained from Eq. (8.33) is a wrapped phase, but it can be unwrapped into a continuous phase by the phase unwrapping algorithm. Using the continuous phase obtained, the out-of-plane displacement derivative can be given by

$$\frac{\partial w_o}{\partial x}=\frac{\delta_u}{2k\Delta}=\frac{\lambda\delta_u}{4\pi\Delta} \tag{8.35}$$

3. Experimental Certification

Four contour fringe patterns of slope with phase-shifting values -3α, $-\alpha$, α and 3α are shown in Fig. 8.8.

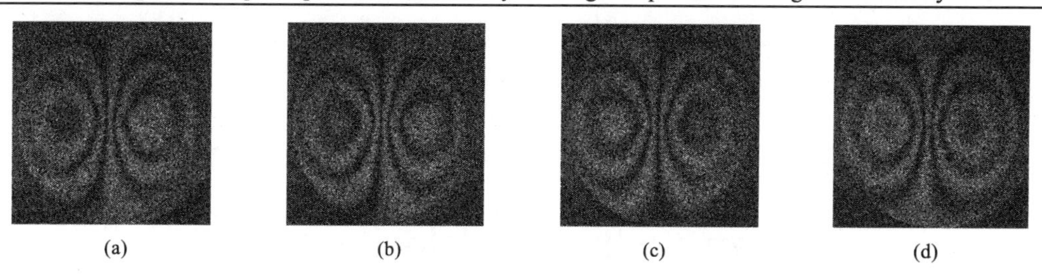

(a) (b) (c) (d)

Fig. 8.8 Slope contour fringes

The phase maps of slope are shown in Fig. 8.9. Fig. 8.9(a) is the contour fringe pattern of slope with zero phase-shifting value; Fig. 8.9(b) is the wrapped phase map of the $0 \sim 2\pi$ range; Fig. 8.9(c) is the continuous phase map.

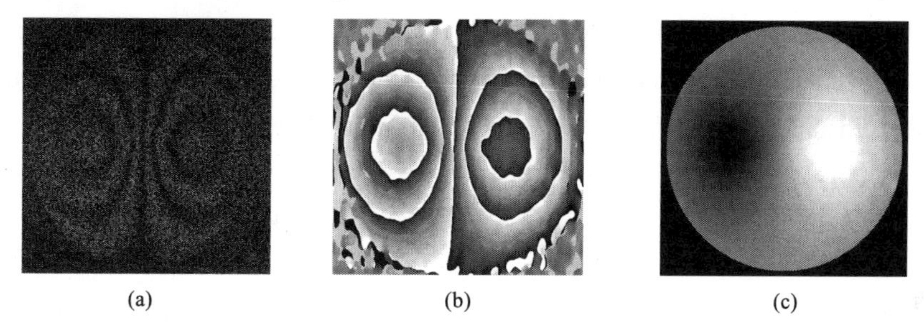

(a) (b) (c)

Fig. 8.9 Phase maps for slope

Chapter 9 Digital Image Correlation and Particle Image Velocimetry

Digital speckle photography is a digital version of speckle photography. In digital speckle photography, specklegrams recorded by a CCD can be stored in different frames. This allows employing the correlation algorithm in digital speckle photography. When the correlation algorithm is utilized, digital speckle photography is usually called digital image correlation in mechanics of solids and particle image velocimetry in mechanics of fluids.

9.1 Digital Image Correlation

Digital image correlation refers to an optical method that employs the correlation property of two displaced speckle fields recorded before and after deformation for measuring the displacement or deformation of deformable objects by determining the extreme position of correlation coefficient.

9.1.1 Image Correlation Principle

Speckles are randomly distributed, i.e., the speckle distribution around any point in a speckle field is different from that around other points, therefore an arbitrary sub-area around a certain point in the speckle field can be used as a carrier to determine displacement and deformation at that point based on movement or change of the sub-area.

Assume that when the object is displaced or deformed, the point to be measured on the specklegram is moved from $P(x,y)$ to $P(x',y')$ and the sub-area around $P(x,y)$ is changed from $\Delta S(x,y)$ into $\Delta S(x',y')$, as shown in Fig. 9.1. Because $\Delta S(x,y)$ and $\Delta S(x',y')$ have the highest similarity, the correlation coefficient between $\Delta S(x,y)$ and $\Delta S(x',y')$ will reach a maximum value. According to the peak position of correlation coefficient, the sub-area displacement can be determined.

Assuming that the sub-area center is moved from $P(x,y)$ to $P(x',y')$, then we have

$$x' = x + u(x,y), \quad y' = y + v(x,y) \tag{9.1}$$

where $u(x,y)$ and $v(x,y)$ are the displacement components of the sub-area center, respectively along the x and y directions.

Consider $Q(x+\Delta x, y+\Delta y)$ adjacent to $P(x,y)$ in the sub-area with Δx and Δy being the distances of these two points before deformation along the x and y directions, respectively. Assume that $Q(x+\Delta x, y+\Delta y)$ is moved to $Q(x'+\Delta x', y'+\Delta y')$ after deformation, in which $\Delta x'$ and $\Delta y'$ can be expressed as

$$\Delta x' = \Delta x + \Delta u(x,y), \quad \Delta y' = \Delta y + \Delta v(x,y) \tag{9.2}$$

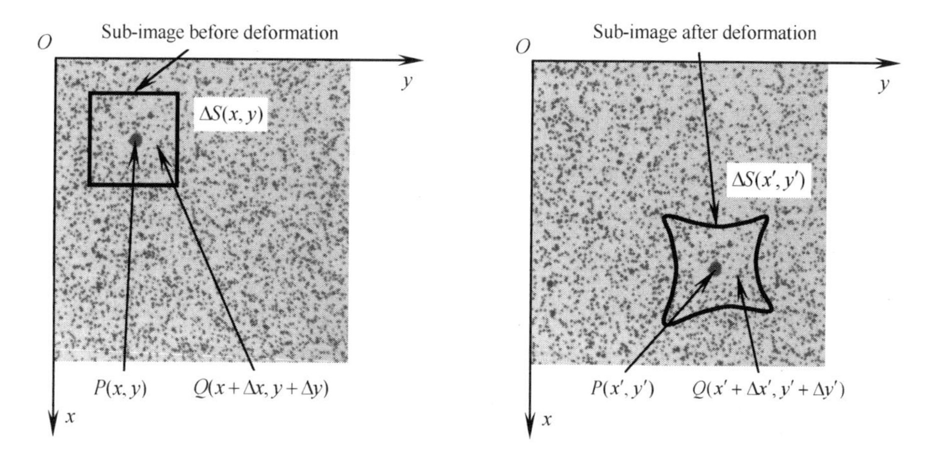

Fig. 9.1 Image correlation principle

where $\Delta u(x,y)$ and $\Delta v(x,y)$ can be expressed, respectively, as

$$\Delta u(x,y) = \frac{\partial u(x,y)}{\partial x}\Delta x + \frac{\partial u(x,y)}{\partial y}\Delta y, \quad \Delta v(x,y) = \frac{\partial v(x,y)}{\partial x}\Delta x + \frac{\partial v(x,y)}{\partial y}\Delta y \tag{9.3}$$

Therefore, when $Q(x+\Delta x, y+\Delta y)$ is moved to $Q(x'+\Delta x', y'+\Delta y')$, two displacement components at $Q(x+\Delta x, y+\Delta y)$ can be expressed as

$$\begin{aligned} u(x+\Delta x) &= (x'+\Delta x') - (x+\Delta x) = u(x,y) + \frac{\partial u(x,y)}{\partial x}\Delta x + \frac{\partial u(x,y)}{\partial y}\Delta y \\ v(x+\Delta x) &= (y'+\Delta y') - (y+\Delta y) = v(x,y) + \frac{\partial v(x,y)}{\partial x}\Delta x + \frac{\partial v(x,y)}{\partial y}\Delta y \end{aligned} \tag{9.4}$$

Clearly, the displacement of any point within the sub-area can be represented by two displacement components u and v, and four displacement derivatives $\partial u/\partial x$, $\partial u/\partial y$, $\partial v/\partial x$ and $\partial v/\partial y$.

9.1.2 Image Correlation Algorithm

1. Auto-Correlation Method

For a high-speed moving object, two instantaneous speckle fields are usually required to be stored into the same frame; thus the auto-correlation method needs to be utilized to determine displacement distribution of these two speckle fields. When various points on a specklegram have different displacements from each other, this specklegram needs to be divided into multiple sub-areas and all speckles within each sub-area have approximately the same displacement. Therefore, the intensity distribution for an arbitrary sub-area can be expressed as

$$I(m,n) = I_1(m,n) + I_2(m,n) \tag{9.5}$$

where $m=0,1,2,\cdots,K-1$ and $n=0,1,2,\cdots,L-1$ with $K\times L$ being the sub-area size. By using the auto-correlation method, the correlation coefficient distribution of the sub-area can be expressed as

$$R=(I-<I>)\circ(I-<I>) \tag{9.6}$$

where $<\cdot>$ represents the ensemble average, $\circ$ represents the correlation operation. Since the Fourier transform of the convolution of two functions is equal to the product of the Fourier transforms of these two functions, Eq. (9.6) can also be expressed as

$$R=\mathrm{IFT}\{\mathrm{FT}\{I\}\mathrm{FT}\{\mathrm{RPI}\{I^*\}\}\} \tag{9.7}$$

where $\mathrm{FT}\{\cdot\}$ and $\mathrm{IFT}\{\cdot\}$ represent respectively the Fourier transform and the inverse Fourier transform, RPI represents the image being rotated by π, and $*$ represents the complex conjugate. In practical applications, correlation operation is usually performed by the fast Fourier transform in the frequency domain, since it is a much faster method than directly computing correlation coefficients in the space domain.

The result obtained from the auto-correlation method is shown in Fig. 9.2. The double-exposed specklegram is shown in Fig. 9.2(a), and the correlation coefficient distribution is shown in Fig. 9.2(b).

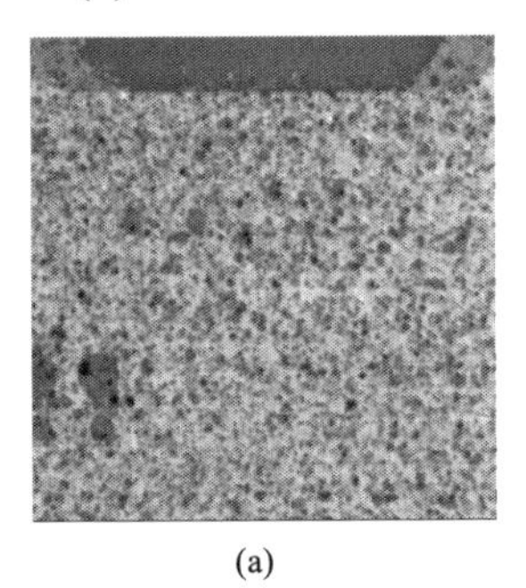
(a)

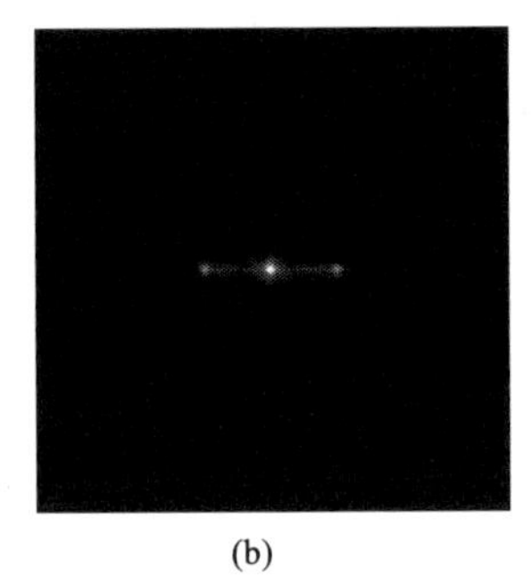
(b)

Fig. 9.2 Auto-correlation result

It can be seen from Fig. 9.2 that, when the auto-correlation method is utilized, the correlation coefficient distribution will have 3 peaks. The distance between the central peak and each of the peaks on both sides is the displacement magnitude of the sub-area center, but the displacement direction cannot be determined.

2. Cross-Correlation Method

If possible, specklegrams before and after deformation are usually stored separately and the displacement of speckles can be determined by the cross-correlation method. Similarly, when the displacements of various points on the specklegram are not different from each other, this specklegram is also needed to be divided into multiple sub-areas so that all speckles in the same sub-area have approximately the same displacement.

Assume that the intensity distributions of the sub-area before and after deformation are respectively represented by $I_1(m,n)$ and $I_2(m,n)$, where $m=0,1,2,\cdots,K-1$, $n=0,1,2,\cdots,L-1$.

From the cross-correlation method, the correlation coefficient of the sub-area before and after deformation can be expressed as

$$R = (I_2 - \langle I_2 \rangle) \circ (I_1 - \langle I_1 \rangle) \tag{9.8}$$

Similarly, Eq. (9.8) can also be expressed as

$$R = \mathrm{IFT}\{\mathrm{FT}\{I_2\}\mathrm{FT}\{\mathrm{RPI}\{I_1{}^*\}\}\} \tag{9.9}$$

The result obtained from the cross-correlation method is shown in Fig. 9.3. Two specklegrams before and after deformation are shown in Fig. 9.3(a) and Fig. 9.3(b), and the correlation coefficient distribution is shown in Fig. 9.3(c). It can be seen from Fig. 9.3 that, when the cross-correlation method is utilized, the correlation coefficient distribution has only one peak. The distance and direction of the peak with respect to the sub-area center are the displacement magnitude and direction of the sub-area center.

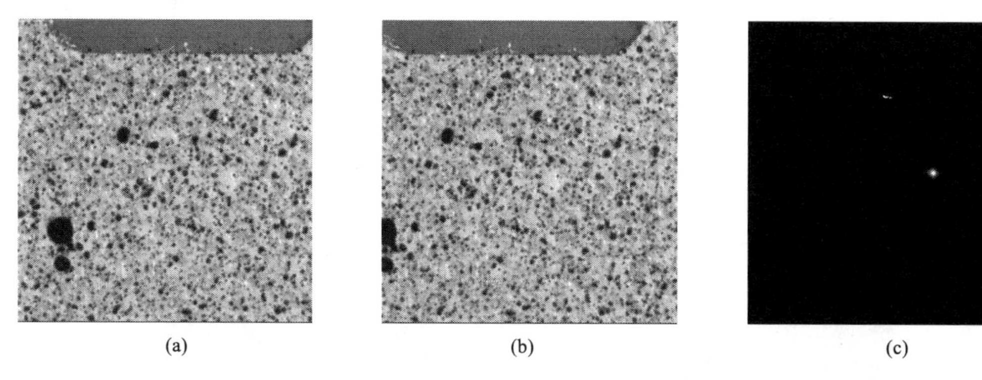

(a) (b) (c)

Fig. 9.3 Cross-correlation result

9.1.3 Image Correlation System

The image correlation system is shown in Fig. 9.4. White light (or laser) can be used to illuminate the specimen. In order to make the light field on the specimen surface be uniformly distributed, symmetrical light sources are usually utilized in the system.

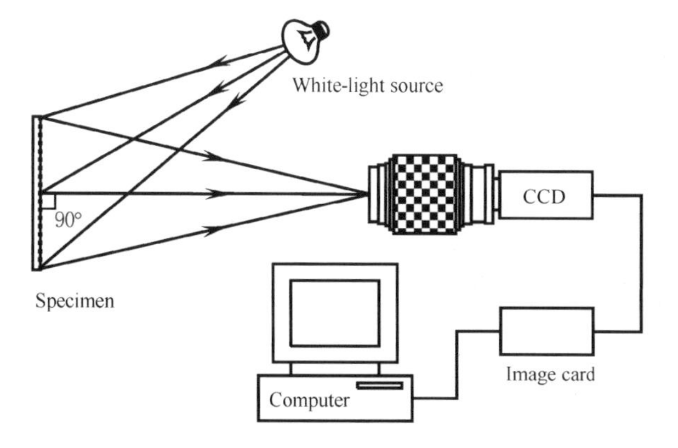

Fig. 9.4 Image correlation system

9.1.4 Image Correlation Experiment

Two white-light digital specklegrams before and after deformation are shown in Fig. 9.5.

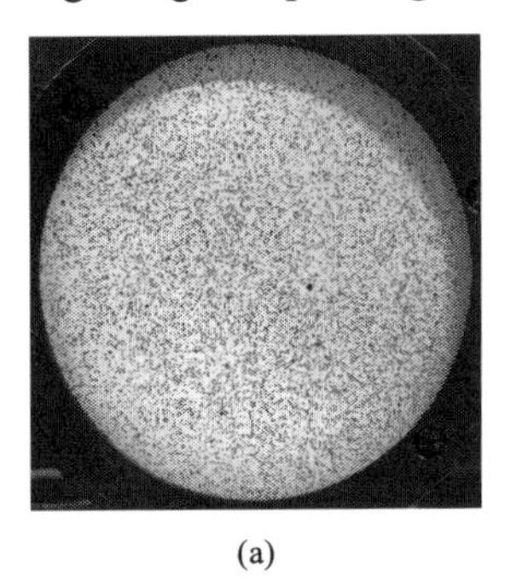

(a)

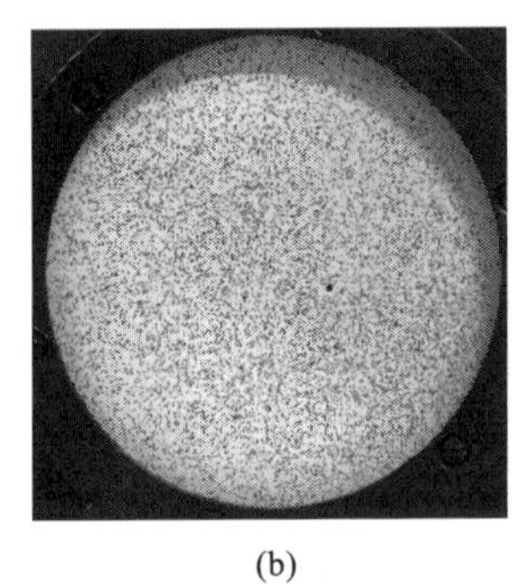

(b)

Fig. 9.5　White-light digital specklegrams

The displacement components obtained by using the digital image correlation method are shown in Fig. 9.6. Fig. 9.6(a) is the vertical component (positive to the downward), and Fig. 9.6 (b) is the horizontal component (positive to the right).

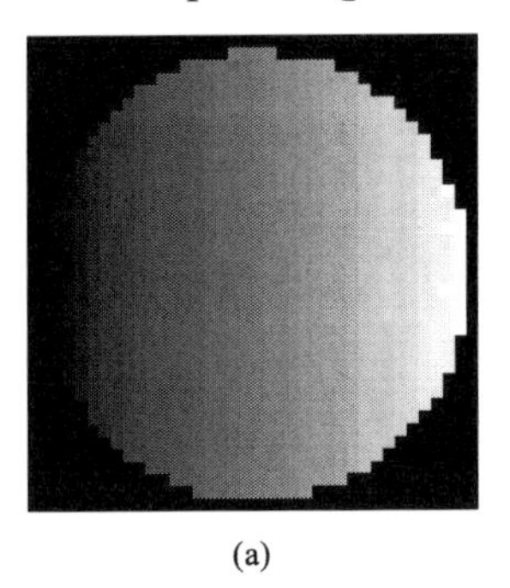

(a)

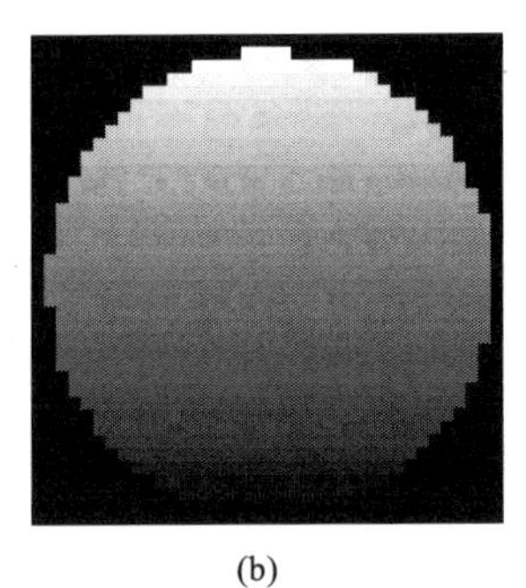

(b)

Fig. 9.6　Displacement components

The magnitude and direction of displacement are shown in Fig. 9.7. Fig. 9.7(a) represents the magnitude and Fig. 9.7(b) the direction.

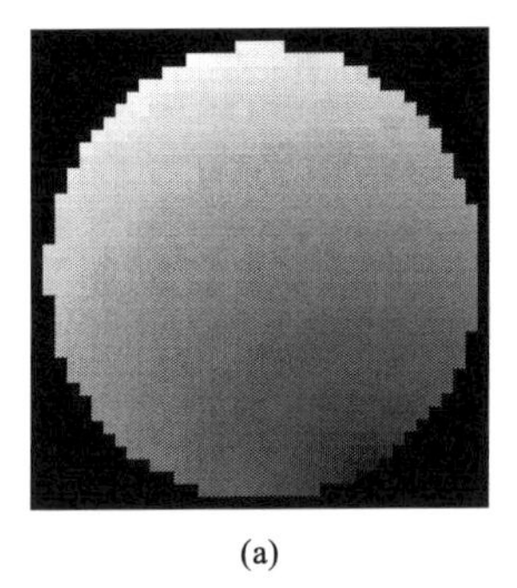

(a)

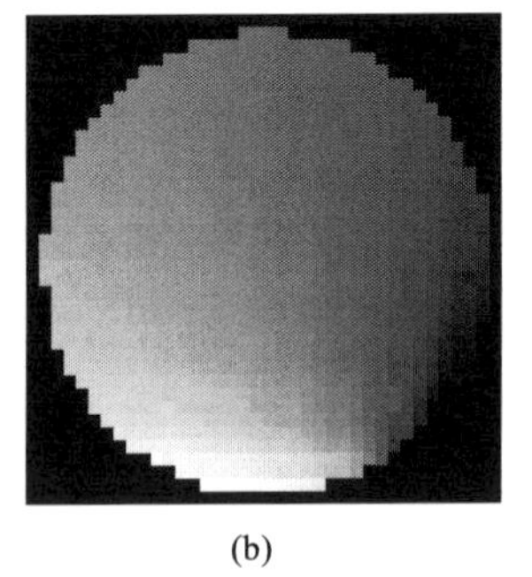

(b)

Fig. 9.7　Magnitude and direction of displacement

9.2 Particle Image Velocimetry

Particle image velocimetry is an optical method that is used for measuring instantaneous velocity in mechanics of fluids. The fluid is seeded with trace particles which are assumed to faithfully follow the flow. The fluid with entrained particles is illuminated so that particles are visible. The velocity distribution of the flow field can be obtained by measuring the movement of the trace particles seeded in the flow field.

Particle image velocimetry, proposed based on speckle photography, is the high-precision, non-contact, full-field technique for measuring flow velocity. Therefore, it has been widely applied to measurement of flow fields.

9.2.1 Image Velocimetry Principle

As shown in Fig. 9.8, assuming that, at the time t and $t+\Delta t$, the trace particle P in the flow field is located at $[x(t), y(t)]$ and $[x(t+\Delta t), y(t+\Delta t)]$, then the displacement components, respectively in the x and y directions, of the trace particle in the time interval Δt can be expressed as

$$u(t) = x(t+\Delta t) - x(t),\ v(t) = y(t+\Delta t) - y(t) \tag{9.10}$$

The corresponding velocity components, respectively in the x and y directions, of the trace particle P in the time interval Δt can be expressed as

$$v_x = \frac{u(t)}{\Delta t},\quad v_y = \frac{v(t)}{\Delta t} \tag{9.11}$$

The time interval Δt is usually very short, thus the velocity components obtained from Eq. (9.11) can be regarded as instantaneous components at the time t.

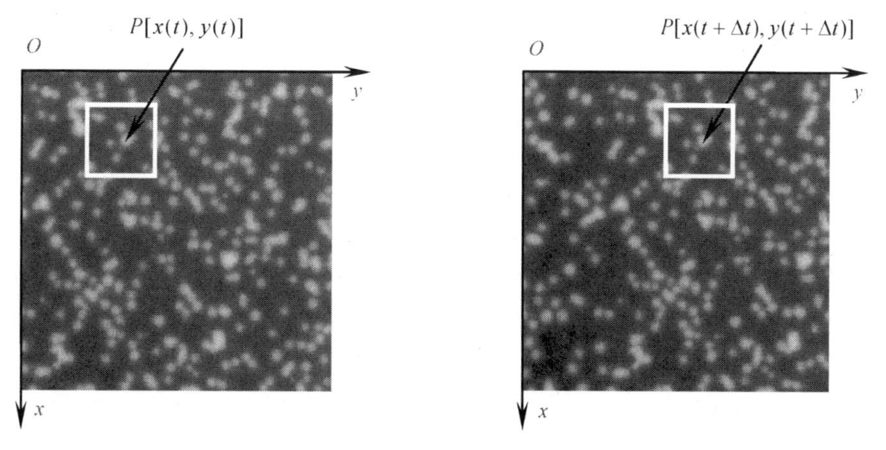

Fig. 9.8 Image velocimetry principle

9.2.2 Image Velocimetry Algorithm

1. Auto-Correlation Method

When the particle images of two different times are stored into the same frame, then the intensity distribution of the sub-area can be expressed as

$$I(m,n)=I_1(m,n)+I_2(m,n) \tag{9.12}$$

where $m=0,1,2,\cdots,K-1$ and $n=0,1,2,\cdots,L-1$ with $K\times L$ being the sub-area size, $I_1(m,n)$ and $I_2(m,n)$ are the intensity distributions at two different times. When the sub-area is small enough, all the particles in the sub-area can be considered to have the same displacement, and we have

$$I_2(m,n)=I_1(m-u,n-v) \tag{9.13}$$

where $u=u(m,n)$ and $v=v(m,n)$ are the displacement components of the particles in the sub-area in the x and y directions. When Eq. (9.12) is subject to the Fourier transform, we obtain

$$A(f_x,f_y)=\mathrm{FT}\{I(m,n)\}=\mathrm{FT}\{I_1(m,n)\}+\mathrm{FT}\{I_2(m,n)\} \tag{9.14}$$

where f_x and f_y are the discrete frequency coordinates along the x and y directions on the transformation plane. By using the shift theorem of the Fourier transform, we have

$$\mathrm{FT}\{I_2(m,n)\}=\mathrm{FT}\{I_1(m-u,n-v)\}=\mathrm{FT}\{I_1(m,n)\}\exp[-\mathrm{i}2\pi(uf_x+vf_y)] \tag{9.15}$$

And Eq. (9.14) can be rewritten by

$$A(f_x,f_y)=\mathrm{FT}\{I_1(m,n)\}\{1+\exp[-\mathrm{i}2\pi(uf_x+vf_y)]\} \tag{9.16}$$

Therefore, the intensity distribution on the transformation plane can be expressed as

$$I(f_x,f_y)=|A(f_x,f_y)|^2=2|\mathrm{FT}\{I_1(m,n)\}|^2\ \{1+\cos[2\pi(uf_x+vf_y)]\} \tag{9.17}$$

Obviously, cosine interference fringes can be observed on the transformation plane.

When Eq. (9.16) is subjected to the auto-correlation operation or Eq. (9.17) is subjected to the Fourier transform, the correlation coefficient distribution of the sub-area can be written by

$$\begin{aligned} R(m,n)=\mathrm{FT}\{|A(f_x,f_y)|^2\}=2\mathrm{FT}\{|\mathrm{FT}\{I_1(m,n)\}|^2\}+\mathrm{FT}\{|\mathrm{FT}\{I_1(m+v,m+v)\}|^2\} \\ +\mathrm{FT}\{|\mathrm{FT}\{I_1(m-u,m-v)\}|^2\} \end{aligned} \tag{9.18}$$

It is clear that the correlation coefficient distribution of the sub-area has 3 peaks. Therefore, the measurement of the sub-area displacement is equivalent to the determination of the distance between the central peak and each of the peaks on both sides, but the displacement direction can not be determined by the auto-correlation method.

The result obtained from the auto-correlation method is shown in Fig. 9.9. The double-exposed specklegram is shown in Fig. 9.9(a), and the correlation coefficient distribution is shown in Fig. 9.9(b).

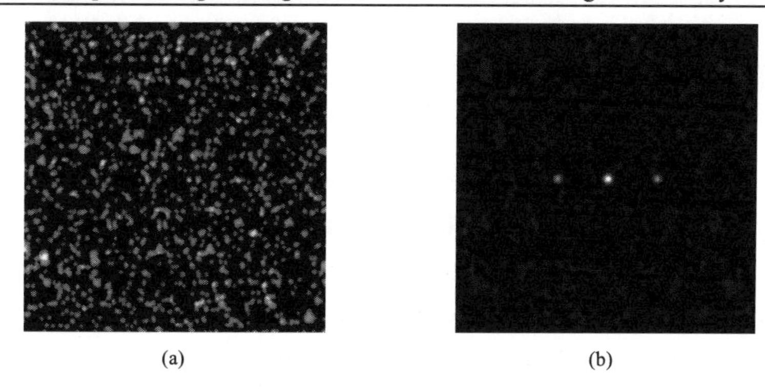

(a) (b)

Fig. 9.9 Auto-correlation result

2. Cross-Correlation Method

Assuming that the intensity distributions of particle patterns recorded at two different times are respectively represented by $I_1(m,n)$ and $I_2(m,n)$, and that all the particles in the sub-area have the same displacement, then we obtain

$$I_2(m,n) = I_1(m-u, n-v) \tag{9.19}$$

If the two sub-areas are subjected to the Fourier transform, then we obtain

$$A_1(f_x, f_y) = \mathrm{FT}\{I_1(m,n)\}, \quad A_2(f_x, f_y) = \mathrm{FT}\{I_1(m,n)\}\exp[-\mathrm{i}2\pi(uf_x + vf_y)] \tag{9.20}$$

When Eq. (9.20) is subjected to the cross-correlation operation, the correlation coefficient distribution of the sub-area can be given by

$$R(m,n) = \mathrm{FT}\{[A_1(f_x, f_y)][A_2(f_x, f_y)]^*\} = \mathrm{FT}\{|\mathrm{FT}\{I_1(m-u, n-v)\}|^2\} \tag{9.21}$$

It is obvious that, when using the cross-correlation method, the correlation coefficient distribution of the sub-area has only 1 peak. The distance and direction of the peak with respect to the sub-area center is the displacement of the sub-area.

The result obtained from the cross-correlation method is shown in Fig. 9.10. Two specklegrams recorded before and after deformation are shown in Fig. 9.10(a) and Fig. 9.10(b), and the correlation coefficient distribution is shown in Fig. 9.10(c).

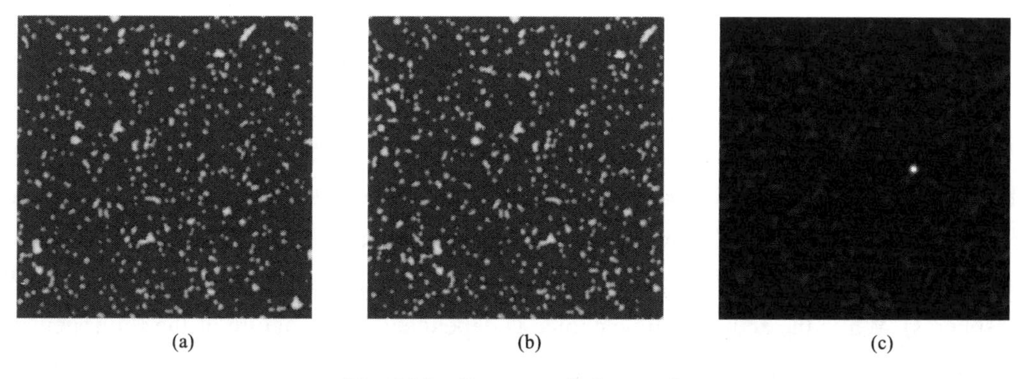

(a) (b) (c)

Fig. 9.10 Cross-correlation result

9.2.3 Image Velocimetry System

The image velocimetry system is shown in Fig. 9.11. A pulsed laser beam emitted from a laser passes through a cylindrical lens to form a laser sheet, which is used to illuminate the flow field to be studied. When the laser sheet illuminating the trace particles in the flow field to be measured is scattered from these trace particles, particle images are recorded by a CCD camera in the direction perpendicular to the plane containing the laser sheet. After two or more times of exposure, the particle images at different times are stored separately into a computer. Using the auto-correlation or cross-correlation operation, the velocity distribution of the flow field can be calculated according to the sub-area displacement produced in the known time interval.

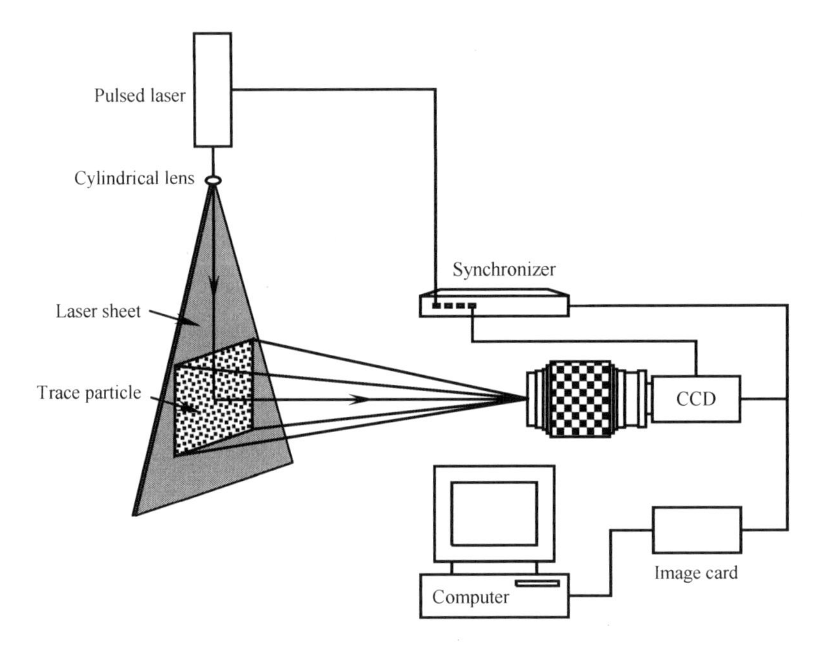

Fig. 9.11　Image velocimetry system

A PIV system mainly includes trace particle, laser sheet, synchronizer, image capturing and processing system, etc.

1. Trace Particles

The trace particles are an inherently critical component of the PIV system. Depending on the fluid under investigation, the particles must be able to match the fluid properties reasonably well. Otherwise they will not follow the flow satisfactorily enough. Ideal particles will have the same density as the fluid system being used, and be spherical (these particles are called microspheres). While the actual particle choice is dependent on the nature of the fluid, generally for macro PIV investigations they are glass beads, polystyrene, polyethylene, aluminum flakes or oil droplets (if the fluid under investigation is a gas). Refractive index for the trace particles should be different from the fluid, so that the laser sheet incident on the fluid will be scattered from the particles and towards the camera.

The particles are typically of a diameter in the order of 10 to 100 micrometers. As for sizing, the particles should be small enough so that response time of the particles to the motion of the fluid is reasonably short to accurately follow the flow, yet large enough to scatter a significant quantity of the incident laser light. The ability of the particles to follow the flow of fluid is inversely proportional to the difference in density between the particles and the fluid, and also inversely proportional to the square of their diameter. The scattered light from the particles is proportional to the square of the diameter of particle. Thus the particle size needs to be balanced to scatter enough light to accurately visualize all particles within the laser sheet plane, but small enough to accurately follow the flow.

2. Laser Sheet

For macro PIV setups, lasers are predominant due to their ability to produce high-power light beams with short pulse durations. This yields short exposure times for each frame. Nd: YAG lasers, commonly used in PIV setups, emit primarily at 1064nm wavelength and its harmonics (532, 266, etc.) For safety reasons, the laser emission is typically band-pass filtered to isolate the 532nm harmonics (this is green light, the only harmonic able to be seen by the naked eye). A fiber optic cable or liquid light guide might be used to direct the laser light to the experimental setup.

The system consisting of a spherical lens and cylindrical lens will produce the laser sheet. The cylindrical lens expands the laser into a plane while the spherical lens compresses the plane into a thin sheet. This is critical as the PIV technique cannot generally measure motion normal to the laser sheet and so ideally this is eliminated by maintaining an entirely 2-dimensional laser sheet. The minimum thickness is on the order of the wavelength of the laser light and occurs at a finite distance from the optical setup (the focal point of the spherical lens). This is the ideal location to place the analysis area of the experiment.

3. Synchronizer

The synchronizer acts as an external trigger for both the camera(s) and the laser. Controlled by a computer, the synchronizer can dictate the timing of each frame of the CCD camera's sequence in conjunction with the firing of the laser to within 1ns precision. Thus the time between each pulse of the laser and the placement of the laser shot in reference to the camera's timing can be accurately controlled. Stand-alone electronic synchronizers, called digital delay generators, offer variable resolution timing from as low as 250ps to as high as several ms.

4. Image Capturing and Processing System

The synchronizer controls the timing between image exposures and also permits image pairs to be acquired at various times along the flow. For accurate PIV analysis, it is ideal that the region of the flow that is of interest should display an average particle displacement of about 8 pixels. The scattered light from each particle should be in the region of 2 to 4 pixels across on the image.

To perform PIV analysis on the flow, two exposures of laser light are required upon the camera from the flow. Originally, with the inability of cameras to capture multiple frames at high speeds, both exposures were captured on the same frame and this single frame was used to determine the flow. A process called auto-correlation was used for this analysis. However, as a result of auto-correlation the direction of the flow becomes unclear, as it is not clear which particle spots are from the first pulse and which are from the second pulse. Faster digital cameras using CCD or CMOS chips were developed since then that can capture two frames at high speed with a few hundred ns difference between the frames. This has allowed each exposure to be isolated on its own frame for more accurate cross-correlation analysis.

The frames are split into a large number of sub-areas. It is then possible to calculate a displacement vector for each sub-area with help of the auto-correlation or cross-correlation techniques. This is converted to a velocity using the time between laser shots and the physical size of each pixel on the camera. The size of the sub-area should be chosen to have at least 6 particles per sub-area on average.

Reference

Born M, Wolf E. 1999. Principles of Optics. 7th ed. Cambridge: Cambridge University Press.

Dainty J C. 1984. Laser Speckle and Related Phenomena. 2nd ed. Berlin: Springer-Verlag.

Erf R K. 1978. Speckle Metrology. London: Academic Press.

Ghiglia D C, Pritt M D. 1998. Two-Dimentional Phase Unwrapping: Theory, Algorithms, and Software. New York: John Wiley & Sons.

Goodman J W. 1968. Introduction to Fourier Optics. San Francisco: McGraw-Hill.

Goodman J W. 1985. Statistical Optics. New York: John Wiley & Sons.

Jones R, Wykes C. 1983. Holographic and Speckle Interferometry: A discussion of the theory, practice and application of the techniques. Cambridge: Cambridge University Press.

Radtogi P K. 1997. Optical Measurement Techniques and Applications. Norwood: Artech House.

Rastogi P K. 2000. Photomechanics. Berlin: Springer-Verlag.

Rastogi P K. 2001. Digital Speckle Pattern Interferometry and Related Techniques. Chichester: John Wiley & Sons.

Rastogi P R, Inaudi D. 2000. Trends in Optical Non-Destructive Testing and Inspection. Amsterdam: Elsevier.

Robison D W, Reid G T. 1993. Interferogram Analysis: Digital Fringe Pattern Measurement Techniques. London: IOP Publishing.

Sirohi R S. 1993. Speckle Metrology. New York: Marcel Dekker.

Sirohi R S. 1999. Optical Methods of Measurement: Wholefield Techniques. New York: Marcel Dekker.

第 1 章　光测力学基础

光测力学是一门实验交叉学科，涉及光学和力学。光测力学通过采用全息干涉、散斑干涉和云纹干涉等光学技术解决变形测量、振动分析和无损检测等力学问题。

1.1　光　　学

光学是与光和视觉有关的物理学分支，主要研究波长位于 X 射线和微波之间电磁辐射的产生、传播和检测。

光学通常刻画光的行为特性。光是电磁辐射，占据电磁频谱部分区间，通常是指能被人眼感知并能形成视觉的可见光。可见光波长处于 400～760nm 范围内，介于红外线和紫外线之间。

红外线和紫外线经常也称为光。红外光是不可见光，波长为 760nm～1mm。紫外光也是不可见光，波长范围为 4～400nm。

1.1.1　几何光学

本身发光或被其他光源照亮后发光的物体称为光源。当光源的尺寸与其辐射距离相比可以忽略不计时，该光源可以看成点光源。在几何光学中，点光源可以抽象成几何点，任何被成像的物体都可看成由无数几何发光点组成。

在几何光学中，光线可以抽象为几何线，其方向表示光波的传播方向。几何光学研究光的传播，实际上就是研究光线的传播。利用光线的概念，可以把复杂的光学成像归结为几何运算问题。目前使用的光学成像系统绝大多数都是根据几何光学原理，利用几何光线概念设计而成。

在各向同性介质中，光线沿着直线传播，这就是光的直线传播定律。该定律可以很好地解释影子、日食和月食等现象。但是，当光在传播过程中遇到很小的不透明屏或通过小孔时，光的传播将偏离直线方向，这就是光的衍射现象。显然，光的直线传播定律只有当光在均匀介质中无阻拦地传播时才成立。

当多束光通过空间某一点时，各光线的传播不受其他光线的影响，称为光的独立传播定律。当两束光汇聚在空间某点时，其作用为简单相加。利用这条定律，我们在研究某一束光的传播时，可以不考虑其他光束的存在。光的独立传播定律只对非相干光束成立。对于相干光束，光的干涉效应将使光的独立传播定律不再成立。

1\. *反射和折射*

当一束光投射到两种透明介质的光滑分界面时，将有一部分光反射回原介质，这部分光称为反射光；另一部分光则通过介质分界面进入第二种介质，这部分光称为折射光，如图 1.1 所示。光线的反射和折射分别满足反射和折射定律。

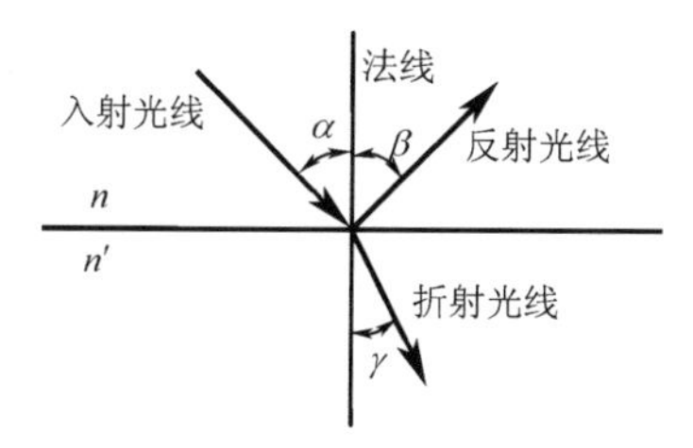

图 1.1 反射和折射

反射定律可表述为入射角和反射角相等，即

$$\alpha = \beta \tag{1.1}$$

式中，α 和 β 分别为入射光线和反射光线与法线之间所形成的锐角。

折射定律可表述为入射角与折射角的正弦之比等于折射光线所在介质与入射光线所在介质的折射率之比，即

$$\frac{\sin\alpha}{\sin\gamma} = \frac{n'}{n} \tag{1.2}$$

式中，α 和 γ 分别为入射光线和折射光线与法线之间所形成的锐角。如果光线由光疏介质入射进入光密介质，即 $n' > n$，则 $\gamma < \alpha$；如果光线由光密介质进入光疏介质，则 $\gamma > \alpha$。

设 $n > n'$ 和 $\alpha = \arcsin\left(\frac{n'}{n}\right)$，则 $\gamma = 90°$，即光线将发生全反射。因此，当光线由光密介质进入光疏介质时，发生全反射的条件可表示为

$$\alpha \geqslant \alpha_{cr} = \arcsin\left(\frac{n'}{n}\right) \tag{1.3}$$

式中，α_{cr} 称为临界角。

2. 透镜

因折射而产生汇聚或发散光线的装置称为透镜。通常存在两类透镜，即引起平行光汇聚的凸透镜和引起平行光发散的凹透镜。薄透镜成像时，焦距、物距和像距满足：

$$\frac{1}{a} + \frac{1}{b} = \frac{1}{f} \tag{1.4}$$

式中，f 为透镜焦距；a 和 b 分别为成像物距和像距。物和像在透镜同侧时 b 为负，在异侧时 b 为正。对于凹透镜，焦距 f 取负值。

入射的平行光线通过凸透镜聚焦将在透镜异侧一倍焦距处形成倒立的实像。有限距离处物体发出的光线将会在大于一倍焦距处聚焦，物体离透镜越近，实像离透镜越远。入射的平行光线通过凹透镜后将会发散，这些光线似乎来源于透镜同侧一倍焦距处的正立虚像。有限距离处物体发出的光线将会在透镜同侧小于一倍焦距处形成虚像，物体离透镜越近，虚像离透镜也越近。成像透镜的放大倍数定义为

$$M = -\frac{b}{a} \tag{1.5}$$

式中，给定负号以保证正立像的放大倍数为正，倒立像的放大倍数为负。类似于平面镜，单透镜产生的正立像为虚像，倒立像为实像。

透镜总会受到像差影响，像差会使图像发生失真。像差既可由透镜几何缺陷引起，也可由不同波长的光的不同折射率引起(色差)。

1.1.2 物理光学

光是电磁波，其振动方向和光的传播方向垂直，即光波是横波。光在真空中的传播速度 $c = 3\times10^8$ m/s，在空气中的传播速度近似等于真空中的速度，而在水和玻璃等透明介质中的传播速度要比真空中慢，其速度与波长和频率的关系可表示为 $v = \lambda\nu$，其中，λ 和 ν 分别为光波的波长和频率。光波在传播过程中，在某一时刻其振动相位相同的各点所构成的曲面称为波面。在各向同性介质中，光波沿着波面法线方向传播，因此可以认为光波波面的法线就是几何光学中的光线。

1. *波动方程*

光波是电磁波，因此满足波动方程

$$\nabla^2 E(r,t) - \frac{1}{c^2}\frac{\partial^2 E(r,t)}{\partial t^2} = 0 \tag{1.6}$$

式中，$\nabla^2 = \frac{\partial^2}{\partial x^2} + \frac{\partial^2}{\partial y^2} + \frac{\partial^2}{\partial z^2}$ 为拉普拉斯算子；$E(r,t)$ 为瞬时光场；c 为光速。上述方程的单色解为

$$E(r,t) = A(r)\exp(-\mathrm{i}\omega t) \tag{1.7}$$

式中，$A(r)$ 为光波复振幅；$\mathrm{i} = \sqrt{-1}$ 为虚数单位；ω 为光波圆频率。把式(1.7)代入式(1.6)，得

$$(\nabla^2 + k^2)A(r) = 0 \tag{1.8}$$

式中，$k = \omega/c = 2\pi/\lambda$ 为波数。式(1.8)即为亥姆霍兹方程。

2. *平面波*

波动方程的平面波解为

$$A(x,y,z) = a\exp(\mathrm{i}kz) \tag{1.9}$$

式中，a 为光波振幅。式(1.9)表示沿 z 方向传播的平面波，其强度分布可表示为

$$I(x,y,z) = |A(x,y,z)|^2 = a^2 \tag{1.10}$$

显然，平面波光场中各点的强度相同。

3. *球面波*

波动方程的球面波解为

$$A(r) = \frac{a}{r}\exp(\mathrm{i}kr) \tag{1.11}$$

式中，a 为单位距离处点的光波振幅。式(1.11)表示沿半径向外传播的球面波，其强度分布可表示为

$$I(r)=|A(r)|^2=\frac{a^2}{r^2} \tag{1.12}$$

即球面波光场中各点的强度分布与 r^2 成反比。

式(1.11)的二阶近似可表示为

$$A(x,y,z)=\frac{a}{z}\exp(\mathrm{i}kz)\exp\left[\frac{\mathrm{i}k}{2z}(x^2+y^2)\right] \tag{1.13}$$

式(1.13)表示距离点光源为 z 的平面上点 (x,y,z) 处的复振幅。

4. 柱面波

波动方程的柱面波解为

$$A(r)=\frac{a}{\sqrt{r}}\exp(\mathrm{i}kr) \tag{1.14}$$

式中，a 为单位距离处的光波振幅。式(1.14)表示沿半径向外传播的柱面波，其强度分布为

$$I(r)=|A(r)|^2=\frac{a^2}{r} \tag{1.15}$$

即柱面波光场中各点的强度分布与 r 成反比。

1.2　光 波 干 涉

频率相同、振动方向相同、相位差保持恒定的两列光波相互叠加后将会出现干涉效应。设两列光波复振幅分别表示为

$$A_1=a_1\exp(\mathrm{i}\varphi_1),\quad A_2=a_2\exp(\mathrm{i}\varphi_2) \tag{1.16}$$

式中，a_1 和 a_2 分别为两列光波的振幅；φ_1 和 φ_2 分别为两列光波的相位。两列光波相互干涉后的强度分布为

$$I=|A_1+A_2|^2={a_1}^2+{a_2}^2+2a_1a_2\cos\Delta\varphi \tag{1.17}$$

式中，$\Delta\varphi=\varphi_2-\varphi_1$ 为相位差。式(1.17)也可表示为

$$I=I_1+I_2+2\sqrt{I_1I_2}\cos\Delta\varphi \tag{1.18}$$

式中，I_1 和 I_2 分别为两列光波的强度分布。由此可见，两列光波相互干涉后的强度分布含有余弦干涉条纹。条纹对比度为

$$V=\frac{I_{\max}-I_{\min}}{I_{\max}+I_{\min}}=\frac{2\sqrt{I_1I_2}}{I_1+I_2} \tag{1.19}$$

式中，$I_{\max}$ 和 $I_{\min}$ 分别为最大强度和最小强度。利用式(1.19)，式(1.18)还可表示为

$$I = I_B + I_M \cos\varphi = I_B(1 + V\cos\Delta\varphi) \tag{1.20}$$

式中，$I_B = I_1 + I_2$ 和 $I_M = 2\sqrt{I_1 I_2}$ 分别为背景强度和调制强度。

1.3 光波衍射

光波在传播过程中遇到障碍物时，将产生偏离直线传播路径的衍射效应。平面单色光波通过孔径后将发生衍射效应，如图 1.2 所示。考虑到傍轴近似，观察面上光波复振幅可表示为

$$A(x,y) = \frac{1}{\mathrm{i}\lambda z}\int_{-\infty}^{\infty}\int_{-\infty}^{\infty} A(\xi,\eta)\exp\left[\mathrm{i}k\sqrt{z^2 + (x-\xi)^2 + (y-\eta)^2}\right]\mathrm{d}\xi\mathrm{d}\eta \tag{1.21}$$

式中，λ 为光波波长；$k = 2\pi/\lambda$ 为波数；$A(\xi,\eta)$ 是衍射孔径面上的光波复振幅。

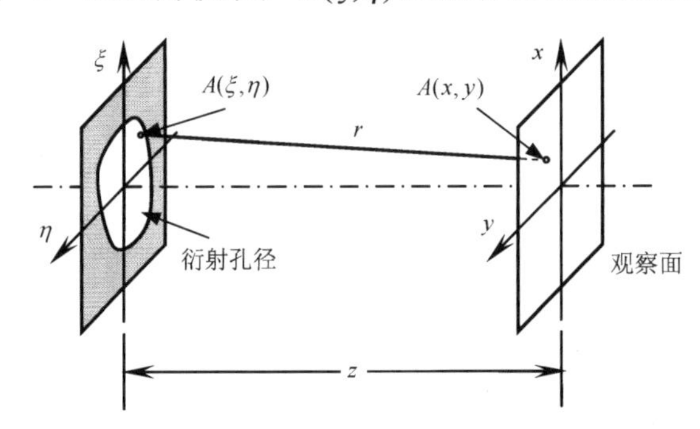

图 1.2　孔径衍射

1. 菲涅尔衍射

在菲涅尔衍射区，有

$$z^3 \gg \frac{\pi}{4\lambda}[(x-\xi)^2 + (y-\eta)^2]^2_{\max} \tag{1.22}$$

则观察面上的光波复振幅可以表示为

$$A(x,y) = \frac{\exp(\mathrm{i}kz)}{\mathrm{i}\lambda z}\exp\left[\mathrm{i}\frac{\pi}{\lambda z}(x^2+y^2)\right]\int_{-\infty}^{\infty}\int_{-\infty}^{\infty} A(\xi,\eta)\exp\left[\mathrm{i}\frac{\pi}{\lambda z}\left(\xi^2+\eta^2\right)\right]\exp\left[-\mathrm{i}\frac{2\pi}{\lambda z}(x\xi+y\eta)\right]\mathrm{d}\xi\mathrm{d}\eta \tag{1.23}$$

2. 夫琅和费衍射

在夫琅和费衍射区，有

$$z \gg \frac{\pi}{\lambda}(\xi^2 + \eta^2)_{\max} \tag{1.24}$$

则观察面上的光波复振幅可以表示为

$$A(x,y) = \frac{\exp(\mathrm{i}kz)}{\mathrm{i}\lambda z}\exp\left[\mathrm{i}\frac{\pi}{\lambda z}(x^2 + y^2)\right]\int_{-\infty}^{\infty}\int_{-\infty}^{\infty} A(\xi,\eta)\exp\left[-\mathrm{i}\frac{2\pi}{\lambda z}(x\xi + y\eta)\right]\mathrm{d}\xi\mathrm{d}\eta \tag{1.25}$$

1.4　光 波 偏 振

普通光源(如太阳和灯泡等)发出的光波在传播过程中，若电场矢量在垂直于传播方向的平面内振动时并不表现出特定的方向性和旋转性，则该光源发出的光波称为非偏振光。如果电场矢量仅在一个特定方向振动，并且在传播过程中振动方向始终保持不变，则该光波称为平面偏振光或线偏振光。如果在光波传播过程中电场矢量振动方向不断围绕传播方向旋转，当电场矢量在垂直于传播方向的平面内绘制出圆形轨迹时称为圆偏振光，当绘制出椭圆轨迹时称为椭圆偏振光。

1.5　光学干涉仪

1. 迈克尔孙干涉仪

图 1.3 所示为迈克尔孙干涉仪。激光器输出的单色光由分光镜分成两束光，其中一束射向固定反射镜，然后反射回分光镜，被分光镜透射的部分光由观察面接收，被分光镜反射的部分光返回到激光器；激光器输出的经分光镜透射的另一束光入射到可动反射镜，经反射后回到分光镜，经分光镜反射的部分光传至观察面，而其余部分光经分光镜透射，返回到激光器。

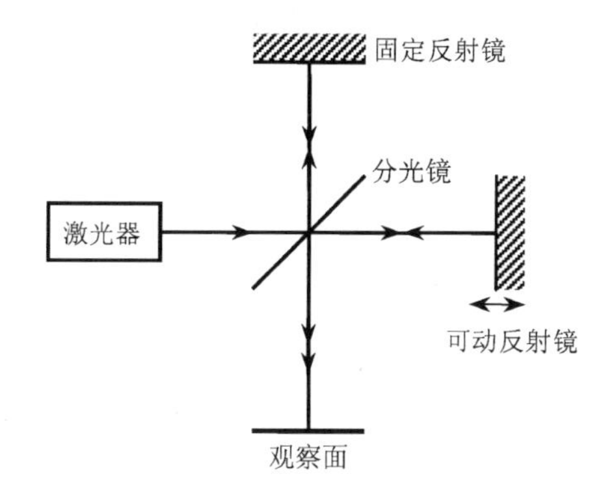

图 1.3　迈克尔孙干涉仪

当两反射镜到分光镜间的光程差小于激光相干长度时，射到观察面上的两光束便产生干涉。

2. 马赫-曾德尔干涉仪

图 1.4 所示为马赫-曾德尔干涉仪。从激光器输出的光束先分后合，两束光由可动反射镜的位移引起相位差，并在观察面上产生干涉。这种干涉仪没有光返回到激光器。此外，从右上方分光镜向上还有另外两束光，一束是上面水平光束的反射部分，另一束是右边垂直光束的透射部分。

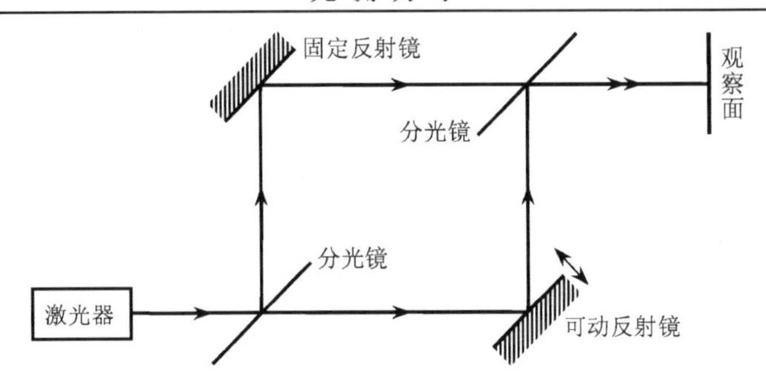

图 1.4　马赫-曾德尔干涉仪

3. 萨奈克干涉仪

图 1.5 为萨奈克干涉仪，它是利用萨奈克效应构成的一种干涉仪。激光器输出的光由分光镜分为反射和透射两部分，这两束光由反射镜的反射形成传播方向相反的闭合光路，然后在分光镜上汇合，被送入观察面中，同时有一部分返回激光器。在这种干涉仪中，任何一块反射镜在垂直于反射表面的方向移动，两光束的光程变化都相同，因此根据双光束干涉原理，在观察面上观察不到干涉强度的变化。但是当把这种干涉仪装在一个可绕垂直于光束平面旋转的平台上且平台以角速度 ω 转动时，根据萨奈克效应，两束传播方向相反的光束到达观察面的相位不同。

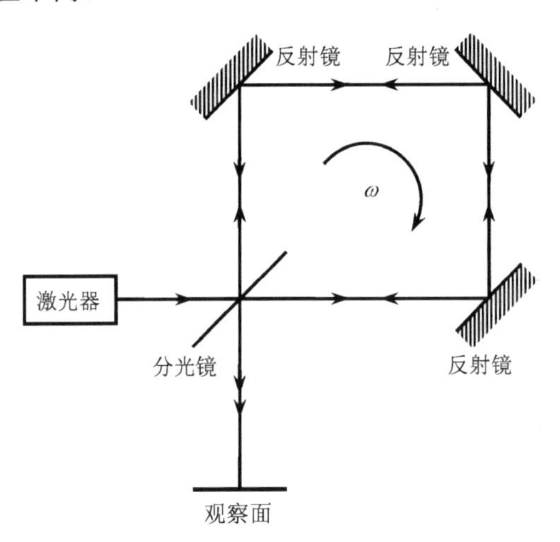

图 1.5　萨奈克干涉仪

4. 法布里-珀罗干涉仪

图 1.6 是法布里-珀罗干涉仪，它由两块平行放置的半透半反镜组成，在两个相对的反射表面镀有反射膜。由激光器输出的光束入射到干涉仪，在两个相对的反射表面作多次往返，透射出去的平行光束由观察面接收。这种干涉仪与前几种干涉仪的根本区别是，前几种干涉仪都是双光束干涉，而法布里-珀罗干涉仪是多光束干涉。

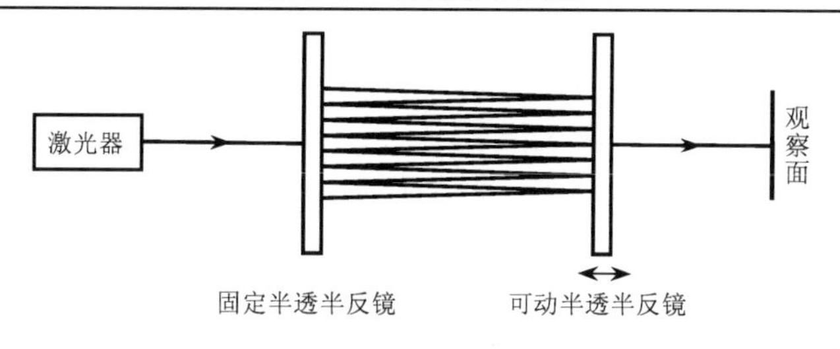

图 1.6　法布里-珀罗干涉仪

1.6　激　光　器

激光器是指能够把具有混合频率的入射光(电磁辐射)转变为具有一个或多个离散频率的高度放大和高度相干的红外线、可见光和紫外线的仪器。

激光器通过其相干性区别于其他光源。空间相干性可以让激光器聚焦到一个微小斑点，并具有很高的强度。空间相干性也可以让激光束在很大距离上始终保持为细束激光(即准直)。激光器也有高度的时间(或纵向)相干性，允许激光器发射窄谱激光，即单色光。时间相干性是指单频偏振波在光束方向较大距离上，其相位具有相关性(相干长度)。

实际上，大部分激光器会产生多模辐射，频率有细微差别(波长)，而不是单一偏振。虽然时间相干性是指单色性，但是仍有发射宽谱光或同时发射不同波长光的激光器。所有这些装置根据它们产生光的方法(即受激辐射)也归类为激光器。

激光器主要应用于对空间或时间相干性要求较高的场合。激光器可用于光盘驱动器、激光打印机和条形码扫描仪，光纤和自由空间光通信，激光外科和皮肤治疗，切削和焊接材料。

由热光源或其他非相干光源产生的光束具有瞬时振幅和相位，随时间和位置随机变化，因而相干长度较短。

第 2 章　全息照相与全息干涉

由于没有高度相干光源，且无法分离孪生像，因此全息在首次提出之后的十多年内并未得到广泛关注。直到发明激光器以及提出离轴法之后，全息才开始有快速发展。从那时起，各种全息方法(如双曝光全息干涉、时间平均全息干涉等)相继提出，各种全息应用(如位移测量、振动分析等)不断涌现。

2.1　全 息 照 相

全息照相利用物体光波和参考光波之间的干涉效应将物体光波的振幅和相位信息以干涉条纹的形式同时记录在照相底片上，照相底片经过显影和定影后变成全息图，然后用再现光波(通常采用记录全息图时的参考光波)照射全息图，通过全息图的衍射效应使物体光波得到再现，进而得到包含物体振幅和相位信息的立体像。因此全息照相不但能记录物体光波的振幅信息，而且能同时记录物体光波的相位信息。

2.1.1　全息记录

全息记录系统如图 2.1 所示。来自激光器的光波经分光镜分光后变成两束光波。其中一束光波经反射镜反射并经扩束镜扩束后照射物体，然后经物体漫反射后照射照相底片。这束来自物体的漫反射光波称为物体光波。另一束光波经反射镜反射并经扩束镜扩束后直接照射照相底片，这束光波称为参考光波。这两列光波相互干涉进而在照相底片上形成干涉图(即全息图)。

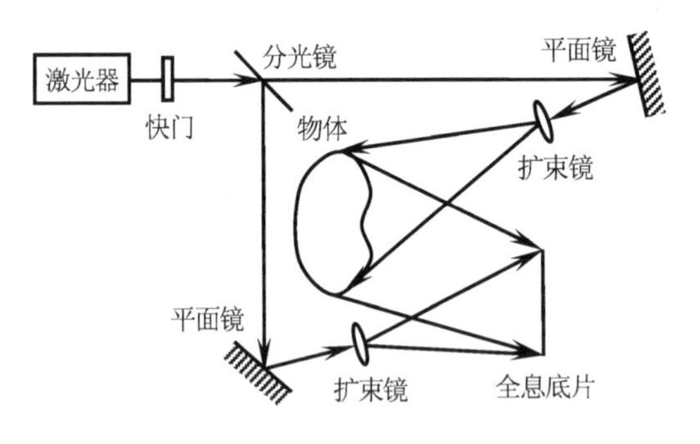

图 2.1　记录系统

设照相底片处物体光波和参考光波的复振幅分别为 $O = a_o\exp(i\varphi_o)$ 和 $R = a_r\exp(i\varphi_r)$，其中 a_o 和 φ_o 分别为物体光波的振幅和相位，a_r 和 φ_r 分别为参考光波的振幅和相位，$i = \sqrt{-1}$ 是虚数单位，则照相底片所记录的强度分布可表示为

$$I = (O+R)(O+R)^* = (a_o^2 + a_r^2) + a_o a_r \exp[i(\varphi_o - \varphi_r)] + a_o a_r \exp[-i(\varphi_o - \varphi_r)] \tag{2.1}$$

式中，* 表示复共轭。利用 $\cos\theta = [\exp(\mathrm{i}\theta) + \exp(-\mathrm{i}\theta)]/2$，式(2.1)还可表示为

$$I = (a_o^2 + a_r^2) + 2a_o a_r \cos(\varphi_o - \varphi_r) \tag{2.2}$$

由此可见，照相底片所记录的强度分布是余弦干涉条纹(不过这些干涉条纹通常很细很密，人眼无法分辨)，因此全息图实际上是一块余弦光栅，当用再现光波照射全息图时，全息图将发生衍射进而产生物体像。

假设曝光时间为 T，则照相底片记录到的曝光量为

$$E = IT = T(a_o^2 + a_r^2) + Ta_o a_r \exp[\mathrm{i}(\varphi_o - \varphi_r)] + Ta_o a_r \exp[-\mathrm{i}(\varphi_o - \varphi_r)] \tag{2.3}$$

在一定曝光量范围内，全息图的振幅透射率与曝光量呈线性关系，取比例常数为 β，那么全息图的振幅透射率可表示为

$$t = \beta E = \beta T(a_o^2 + a_r^2) + \beta Ta_o a_r \exp[\mathrm{i}(\varphi_o - \varphi_r)] + \beta Ta_o a_r \exp[-\mathrm{i}(\varphi_o - \varphi_r)] \tag{2.4}$$

2.1.2　全息再现

全息再现系统如图 2.2 所示。该再现系统与图 2.1 所示的记录系统相同，只是在再现系统中已移走物体，并挡掉物体光波。为了观察物体像，需要把经过显影和定影的照相底片放回原位，用原参考光波照射全息图。

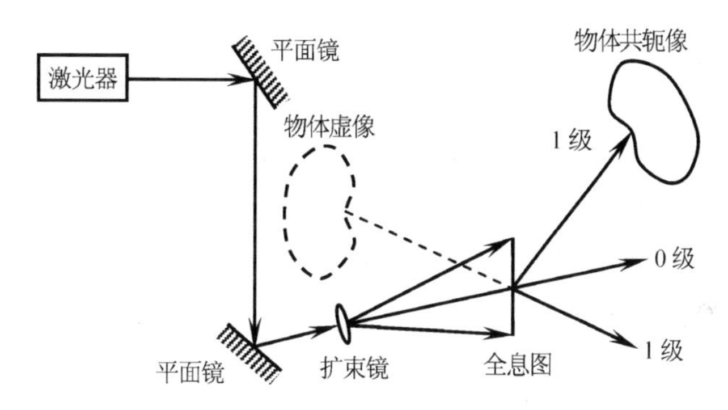

图 2.2　再现系统

当用原参考光波 $R = a_r \exp(\mathrm{i}\varphi_r)$ 照射全息图时，透过全息图的光波复振幅为

$$A = Rt = \beta T(a_o^2 a_r + a_r^3)\exp(\mathrm{i}\varphi_r) + \beta Ta_o a_r^2 \exp(\mathrm{i}\varphi_o) + \beta Ta_o a_r^2 \exp[-\mathrm{i}(\varphi_o - 2\varphi_r)] \tag{2.5}$$

式中，等号右边第一项含有 $\exp(\mathrm{i}\varphi_r)$，表示该项是透过全息图后沿 $\exp(\mathrm{i}\varphi_r)$ 方向传播的 0 级衍射光波；第二项含有 $\exp(\mathrm{i}\varphi_o)$，表示该项是透过全息图后沿 $\exp[\mathrm{i}(\varphi_r - \varphi_r)]$ 方向传播的正 1 级衍射光波，它是物体光波 $\exp(\mathrm{i}\varphi_o)$ 的再现，该项构成了物体虚像，如果这个光波被接收，则可得到与原物完全一样的立体像；第三项含有 $\exp(-\mathrm{i}\varphi_o)\exp(\mathrm{i}2\varphi_r)$，表示该项是沿 $\exp[\mathrm{i}(\varphi_r + \varphi_r)]$ 方向传播的负 1 级衍射光波，它是物体共轭光波 $\exp(-\mathrm{i}\varphi_o)$ 的再现，该项构成了物体共轭实像，如果这个光波被接收，则可得到与原物相位相反的立体像。

上述三个衍射光波沿不同方向传播，彼此相互分离，因此当用原参考光波照射全息图时，透过全息图将有三束光波沿不同方向射出，这就是离轴全息照相。

2.2　全 息 干 涉

全息干涉是基于全息照相而发展起来的高精度非接触全场干涉测量方法，可用于物体的变形测量和振动分析。

2.2.1　相位计算

如图 2.3 所示，设物体变形后，物点从 P 移到 P'，其位移矢量为 $\boldsymbol{d}$。S 和 O 分别表示光源和观察位置，P 和 P' 相对于 S 的位置矢量分别为 $\boldsymbol{r}_0$ 和 $\boldsymbol{r}_0'$；O 相对于 P 和 P' 的位置矢量分别为 $\boldsymbol{r}$ 和 $\boldsymbol{r}'$。因此，物点从 P 移到 P' 后物体光波相位变化可以表示为

$$\delta = k[(\boldsymbol{e}_0 \cdot \boldsymbol{r}_0 + \boldsymbol{e} \cdot \boldsymbol{r}) - (\boldsymbol{e}_0' \cdot \boldsymbol{r}_0' + \boldsymbol{e}' \cdot \boldsymbol{r}')] \tag{2.6}$$

式中，$k = 2\pi/\lambda$ 为物光波数，其中 λ 为波长；$\boldsymbol{e}_0$ 和 $\boldsymbol{e}_0'$ 分别为沿 $\boldsymbol{r}_0$ 和 $\boldsymbol{r}_0'$ 方向的单位矢量；$\boldsymbol{e}$ 和 $\boldsymbol{e}'$ 分别为沿 $\boldsymbol{r}$ 和 $\boldsymbol{r}'$ 方向的单位矢量。

对于小变形，有 $\boldsymbol{e}_0' \approx \boldsymbol{e}_0, \boldsymbol{e}' \approx \boldsymbol{e}$，则式(2.6)可写为

$$\delta = k[\boldsymbol{e}_0 \cdot (\boldsymbol{r}_0 - \boldsymbol{r}_0') + \boldsymbol{e} \cdot (\boldsymbol{r} - \boldsymbol{r}')] \tag{2.7}$$

利用 $\boldsymbol{r}_0 - \boldsymbol{r}_0' = -\boldsymbol{d}, \boldsymbol{r} - \boldsymbol{r}' = \boldsymbol{d}$，可得

$$\delta = k(\boldsymbol{e} - \boldsymbol{e}_0) \cdot \boldsymbol{d} \tag{2.8}$$

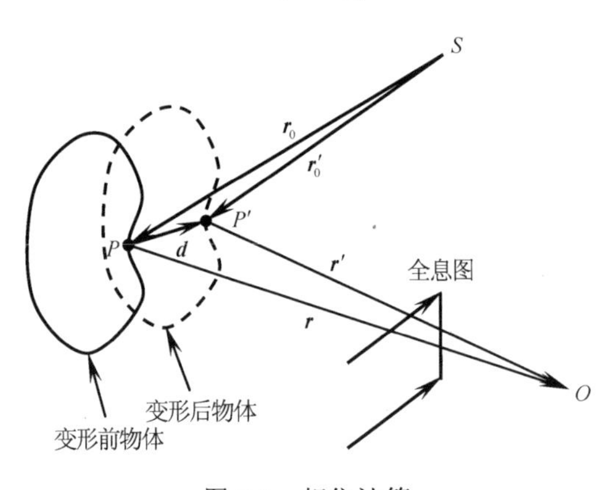

图 2.3　相位计算

2.2.2　双曝光全息干涉

双曝光全息干涉通过两次曝光把对应于物体变形前后的两个不同状态记录于同一张照相底片上。照相底片经过显影和定影处理后，再放回原记录系统进行再现，则对应于物体变形前后的两个物体光波，因相位不同而发生干涉并形成干涉条纹，通过对干涉条纹进行分析，即可实现物体的位移和变形测量。

双曝光全息干涉记录系统如图 2.4 所示。设物体变形前后的物体光波复振幅分别为 $O_1 = a_o \exp(\mathrm{i}\varphi_o)$ 和 $O_2 = a_o \exp[\mathrm{i}(\varphi_o + \delta)]$，其中 a_o 为物体光波振幅(设变形前后振幅不变)；φ_o 和 $(\varphi_o + \delta)$ 分别为变形前后物体光波相位；δ 为因物体变形而引起的物体光波的相位

变化。设参考光波复振幅为 $R = a_\mathrm{r}\exp(\mathrm{i}\varphi_\mathrm{r})$，那么物体变形前后照相底片记录的强度分别为

$$\begin{aligned} I_1 &= (O_1 + R)\cdot(O_1 + R)^* = (a_\mathrm{o}^2 + a_\mathrm{r}^2) + a_\mathrm{o}a_\mathrm{r}\exp[\mathrm{i}(\varphi_\mathrm{o} - \varphi_\mathrm{r})] + a_\mathrm{o}a_\mathrm{r}\exp[-\mathrm{i}(\varphi_\mathrm{o} - \varphi_\mathrm{r})] \\ I_2 &= (O_2 + R)\cdot(O_2 + R)^* = (a_\mathrm{o}^2 + a_\mathrm{r}^2) + a_\mathrm{o}a_\mathrm{r}\exp[\mathrm{i}(\varphi_\mathrm{o} + \delta - \varphi_\mathrm{r})] + a_\mathrm{o}a_\mathrm{r}\exp[-\mathrm{i}(\varphi_\mathrm{o} + \delta - \varphi_\mathrm{r})] \end{aligned} \tag{2.9}$$

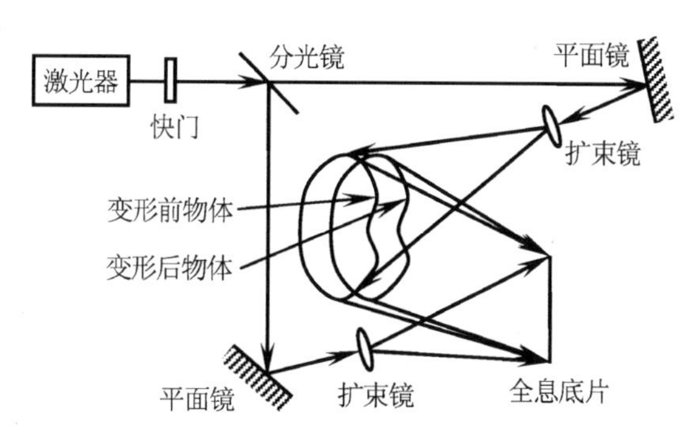

图 2.4　双曝光全息图记录系统

设物体变形前后照相底片的曝光时间均为 T，则照相底片记录到的曝光量可表示为

$$\begin{aligned} E = (I_1 + I_2)T = 2(a_\mathrm{o}^2 + a_\mathrm{r}^2)T &+ a_\mathrm{o}a_\mathrm{r}T\exp[\mathrm{i}(\varphi_\mathrm{o} - \varphi_\mathrm{r})][1 + \exp(\mathrm{i}\delta)] \\ &+ a_\mathrm{o}a_\mathrm{r}T\exp[-\mathrm{i}(\varphi_\mathrm{o} - \varphi_\mathrm{r})][1 + \exp(-\mathrm{i}\delta)] \end{aligned} \tag{2.10}$$

照相底片经显影和定影后，设振幅透射率与曝光量呈线性关系，取比例常数为 β，则双曝光全息图的振幅透射率为

$$\begin{aligned} t = \beta E = 2\beta(a_\mathrm{o}^2 + a_\mathrm{r}^2)T &+ \beta a_\mathrm{o}a_\mathrm{r}T\exp[\mathrm{i}(\varphi_\mathrm{o} - \varphi_\mathrm{r})][1 + \exp(\mathrm{i}\delta)] \\ &+ \beta a_\mathrm{o}a_\mathrm{r}T\exp[-\mathrm{i}(\varphi_\mathrm{o} - \varphi_\mathrm{r})][1 + \exp(-\mathrm{i}\delta)] \end{aligned} \tag{2.11}$$

双曝光全息干涉再现系统如图 2.5 所示。用参考光波 $R = a_\mathrm{r}\exp(\mathrm{i}\varphi_\mathrm{r})$ 照射经显影和定影后的照相底片，则透过全息图的光波复振幅表示为

$$\begin{aligned} A = Rt = 2\beta(a_\mathrm{o}^2a_\mathrm{r} + a_\mathrm{r}^3)T\exp(\mathrm{i}\varphi_\mathrm{r}) &+ \beta a_\mathrm{o}a_\mathrm{r}^2T\exp(\mathrm{i}\varphi_\mathrm{o})[1 + \exp(\mathrm{i}\delta)] \\ &+ \beta a_\mathrm{o}a_\mathrm{r}^2T\exp[-\mathrm{i}(\varphi_\mathrm{o} - 2\varphi_\mathrm{r})][1 + \exp(-\mathrm{i}\delta)] \end{aligned} \tag{2.12}$$

式中，等号右边的第一项是透过全息图后沿参考光波方向的 0 级衍射光波；第二项是透过全息图后沿物体光波方向的正 1 级衍射光波；第三项是物体共轭光波。

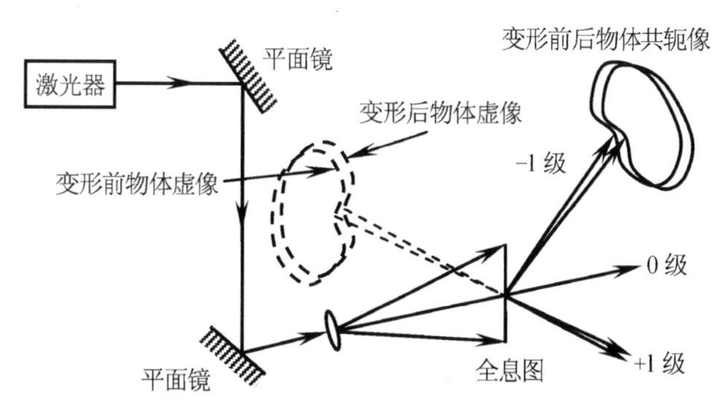

图 2.5　双曝光再现系统

仅考虑含有 $\exp(\mathrm{i}\varphi_{\mathrm{o}})$ 的第二项，则透过全息图的复振幅为

$$A' = \beta a_{\mathrm{o}} a_{\mathrm{r}}^2 T \exp(\mathrm{i}\varphi_{\mathrm{o}})[1+\exp(\mathrm{i}\delta)] \tag{2.13}$$

相应的强度分布为

$$I' = A'A'^* = 2(\beta a_{\mathrm{o}} a_{\mathrm{r}}^2 T)^2 (1+\cos\delta) \tag{2.14}$$

显然，当满足条件

$$\delta = 2n\pi \quad (n = 0, \pm 1, \pm 2, \cdots) \tag{2.15}$$

时，将形成亮纹；当满足条件

$$\delta = (2n+1)\pi \quad (n = 0, \pm 1, \pm 2, \cdots) \tag{2.16}$$

时，将形成暗纹。

把 $\delta = k(\boldsymbol{e}-\boldsymbol{e}_0)\cdot\boldsymbol{d}$ 代入式(2.14)，得

$$I' = 2(\beta T a_{\mathrm{o}} a_{\mathrm{r}}^2)^2 \{1+\cos[k(\boldsymbol{e}-\boldsymbol{e}_0)\cdot\boldsymbol{d}]\} \tag{2.17}$$

当满足条件

$$(\boldsymbol{e}-\boldsymbol{e}_0)\cdot\boldsymbol{d} = n\lambda \quad (n = 0, \pm 1, \pm 2, \cdots) \tag{2.18}$$

时，将形成亮纹；当满足条件

$$(\boldsymbol{e}-\boldsymbol{e}_0)\cdot\boldsymbol{d} = \left(n+\frac{1}{2}\right)\lambda \quad (n = 0, \pm 1, \pm 2, \cdots) \tag{2.19}$$

时，将形成暗纹。

2.2.3　实时全息干涉

实时全息干涉是指先把物体变形前的状态记录在照相底片上，照相底片经过显影和定影处理后精确复位，再用物体光波和参考光波同时照射复位单曝光全息图。如果此时物体发生变形，则可实时观察到干涉条纹，如果物体发生连续变形，也可观察到连续变化的干涉条纹。

设物体变形前的物体光波和参考光波复振幅分别为 $O = a_{\mathrm{o}}\exp(\mathrm{i}\varphi_{\mathrm{o}})$ 和 $R = a_{\mathrm{r}}\exp(\mathrm{i}\varphi_{\mathrm{r}})$，则物体变形前照相底片记录的强度可表示为

$$I = (O+R)(O+R)^* = (a_{\mathrm{o}}^2 + a_{\mathrm{r}}^2) + a_{\mathrm{o}}a_{\mathrm{r}}\exp[\mathrm{i}(\varphi_{\mathrm{o}}-\varphi_{\mathrm{r}})] + a_{\mathrm{o}}a_{\mathrm{r}}\exp[-\mathrm{i}(\varphi_{\mathrm{o}}-\varphi_{\mathrm{r}})] \tag{2.20}$$

设曝光时间为 T，如果振幅透射率与曝光量呈线性关系，设比例常数为 β，则照相底片经显影和定影后的振幅透射率为

$$t = \beta IT = \beta T(a_{\mathrm{o}}^2 + a_{\mathrm{r}}^2) + \beta T a_{\mathrm{o}}a_{\mathrm{r}}\exp[\mathrm{i}(\varphi_{\mathrm{o}}-\varphi_{\mathrm{r}})] + \beta T a_{\mathrm{o}}a_{\mathrm{r}}\exp[-\mathrm{i}(\varphi_{\mathrm{o}}-\varphi_{\mathrm{r}})] \tag{2.21}$$

上述经显影和定影的单曝光全息图精确放回原记录系统进行再现，并用原物体光波和原参考光波同时照射单曝光全息图，如图 2.6 所示。如果此时物体发生变形，则物体变形后的物体光波复振幅可表示为 $O' = a_{\mathrm{o}}\exp[\mathrm{i}(\varphi_{\mathrm{o}}+\delta)]$，其中 δ 为因物体变形而引起的物体光波的相位变化。因此，透过单曝光全息图的光波复振幅可表示为

$$
\begin{aligned}
A=(O'+R)t=&\beta T(a_o^3+a_o a_r^2)\exp[\mathrm{i}(\varphi_o+\delta)]+\beta Ta_o^2a_r\exp[\mathrm{i}(2\varphi_o-\varphi_r+\delta)]\\
&+\beta Ta_o^2a_r\exp[\mathrm{i}(\varphi_r+\delta)]+\beta T(a_o^2a_r+a_r^3)\exp(\mathrm{i}\varphi_r)\\
&+\beta Ta_o a_r^2\exp(\mathrm{i}\varphi_o)+\beta Ta_o a_r^2\exp[-\mathrm{i}(\varphi_o-2\varphi_r)]
\end{aligned}
\tag{2.22}
$$

式中，等号右边的第一项 $\beta T(a_o^3+a_o a_r^2)\exp[\mathrm{i}(\varphi_o+\delta)]$ 和第五项 $\beta Ta_o a_r^2\exp(\mathrm{i}\varphi_o)$ 与物体光波有关，取出这两项，得

$$A'=\beta T\exp(\mathrm{i}\varphi_o)[a_o a_r^2+(a_o^3+a_o a_r^2)\exp(\mathrm{i}\delta)] \tag{2.23}$$

通常 $I_o\ll I_r$，即 $a_o^2\ll a_r^2$，因此式(2.23)简化为

$$A'=\beta Ta_o a_r^2\exp(\mathrm{i}\varphi_o)[1+\exp(\mathrm{i}\delta)] \tag{2.24}$$

相应的强度分布为

$$I'=A'A'^*=2(\beta Ta_o a_r^2)^2(1+\cos\delta) \tag{2.25}$$

将 $\delta=\dfrac{2\pi}{\lambda}(\boldsymbol{e}-\boldsymbol{e}_0)\cdot\boldsymbol{d}$ 代入式(2.25)，可得

$$I'=2(\beta Ta_o a_r^2)^2\{1+\cos[k(\boldsymbol{e}-\boldsymbol{e}_0)\cdot\boldsymbol{d}]\} \tag{2.26}$$

因此，当满足条件

$$(\boldsymbol{e}-\boldsymbol{e}_0)\cdot\boldsymbol{d}=n\lambda\quad(n=0,\pm1,\pm2,\cdots) \tag{2.27}$$

时，将形成亮纹；当满足条件

$$(\boldsymbol{e}-\boldsymbol{e}_0)\cdot\boldsymbol{d}=(n+\tfrac{1}{2})\lambda\quad(n=0,\pm1,\pm2,\cdots) \tag{2.28}$$

时，将形成暗纹。

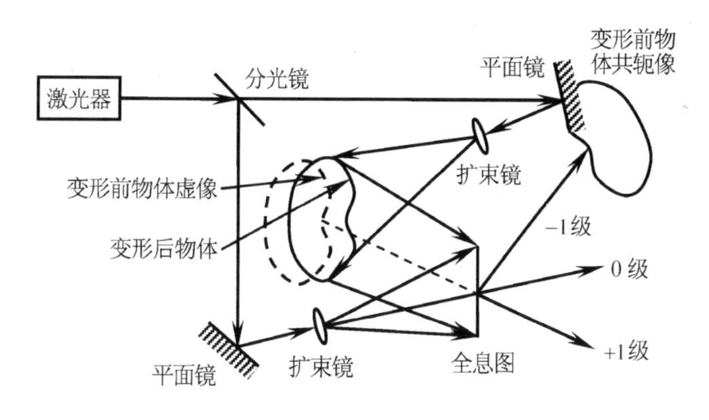

图 2.6 实时再现系统

2.2.4 时间平均全息干涉

时间平均全息干涉是指对稳态振动物体，采用比振动周期长得多的时间进行连续曝光，记录全息图。时间平均法把振动物体在曝光时间内的所有振动状态的物体光波都记录在同一张照相底片上，照相底片经过显影和定影处理后进行再现，此时记录在照相底片上的所有振动状态的物体光波将相互干涉而形成干涉条纹。

设振动物体在任意时刻的物体光波复振幅为 $O = a_\text{o} \exp[\text{i}(\varphi_\text{o} + \delta)]$，$a_\text{o}$ 为物体光波振幅，$(\varphi_\text{o} + \delta)$ 为振动物体在任意时刻的物体光波相位，δ 为物体振动引起的物体光波的相位变化。如果物体做简谐振动，则其振动方程可表示为

$$\boldsymbol{d} = \boldsymbol{A} \sin \omega t \tag{2.29}$$

式中，$\boldsymbol{A}$ 为振动物体的振幅矢量；ω 为振动物体的圆频率或角频率。由此得到任意时刻物体光波的相位变化为

$$\delta = k(\boldsymbol{e} - \boldsymbol{e}_0) \cdot \boldsymbol{d} = k(\boldsymbol{e} - \boldsymbol{e}_0) \cdot \boldsymbol{A} \sin \omega t \tag{2.30}$$

设参考光波复振幅为 $R = a_\text{r} \exp(\text{i}\varphi_\text{r})$。当物体发生振动时任意时刻照相底片记录的强度可表示为

$$I = (O + R)(O + R)^* = (a_\text{o}^2 + a_\text{r}^2) + a_\text{o} a_\text{r} \exp[\text{i}(\varphi_\text{o} - \varphi_\text{r} + \delta)] + a_\text{o} a_\text{r} \exp[-\text{i}(\varphi_\text{o} - \varphi_\text{r} + \delta)] \tag{2.31}$$

设曝光时间为 T，并仅考虑与物体光波有关的项(式(2.31)中等号右边第二项)，则照相底片记录到的曝光量为

$$E = \int_0^T a_\text{o} a_\text{r} \exp[\text{i}(\varphi_\text{o} - \varphi_\text{r} + \delta)] \text{d}t = a_\text{o} a_\text{r} \exp[\text{i}(\varphi_\text{o} - \varphi_\text{r})] \int_0^T \exp[\text{i}k(\boldsymbol{e} - \boldsymbol{e}_0) \cdot \boldsymbol{A} \sin \omega t] \text{d}t \tag{2.32}$$

当 $T >> 2\pi/\omega$ 时，式(2.32)可表示为

$$E = T a_\text{o} a_\text{r} \exp[\text{i}(\varphi_\text{o} - \varphi_\text{r})] J_0[k(\boldsymbol{e} - \boldsymbol{e}_0) \cdot \boldsymbol{A}] \tag{2.33}$$

式中，J_0 为第一类零阶贝塞尔函数。

照相底片经显影和定影后，设振幅透射率与曝光量呈线性关系，取比例常数为 β，则振幅透射率可表示为

$$t = \beta E = \beta T a_\text{o} a_\text{r} \exp[\text{i}(\varphi_\text{o} - \varphi_\text{r})] J_0[k(\boldsymbol{e} - \boldsymbol{e}_0) \cdot \boldsymbol{A}] \tag{2.34}$$

用参考光波 $R = a_\text{r} \exp(\text{i}\varphi_\text{r})$ 照射经显影和定影后的照相底片，则透过全息图的光波复振幅表示为

$$A' = Rt = \beta T a_\text{o} a_\text{r}^2 \exp(\text{i}\varphi_\text{o}) J_0[k(\boldsymbol{e} - \boldsymbol{e}_0) \cdot \boldsymbol{A}] \tag{2.35}$$

相应的强度分布为

$$I' = A' A'^* = (\beta T a_\text{o} a_\text{r}^2)^2 J_0^2[k(\boldsymbol{e} - \boldsymbol{e}_0) \cdot \boldsymbol{A}] \tag{2.36}$$

式(2.36)表明，时间平均法的强度分布与 $J_0^2[k(\boldsymbol{e} - \boldsymbol{e}_0) \cdot \boldsymbol{A}]$ 有关。在 $J_0^2[k(\boldsymbol{e} - \boldsymbol{e}_0) \cdot \boldsymbol{A}]$ 取极大值处将形成亮纹。特别地，当 $J_0^2[k(\boldsymbol{e} - \boldsymbol{e}_0) \cdot \boldsymbol{A}]$ 取最大值时振幅 $A = 0$，即条纹最亮处为振动节线。在 $J_0^2[k(\boldsymbol{e} - \boldsymbol{e}_0) \cdot \boldsymbol{A}]$ 取极小值处将形成暗纹，即当振幅 A 满足条件

$$k(\boldsymbol{e} - \boldsymbol{e}_0) \cdot \boldsymbol{A} = \alpha \quad (\alpha = 2.41,\ 5.52,\ 8.65,\ 11.79,\ 14.98,\ \cdots) \tag{2.37}$$

时将产生暗纹。式中，α 为第一类零阶贝塞尔函数 J_0 的根，即 $J_0(\alpha) = 0$。因此，通过确定条纹级数即可求得振动物体各点的振幅大小。$J_0^2(\alpha)$-α 的分布曲线如图 2.7 所示。

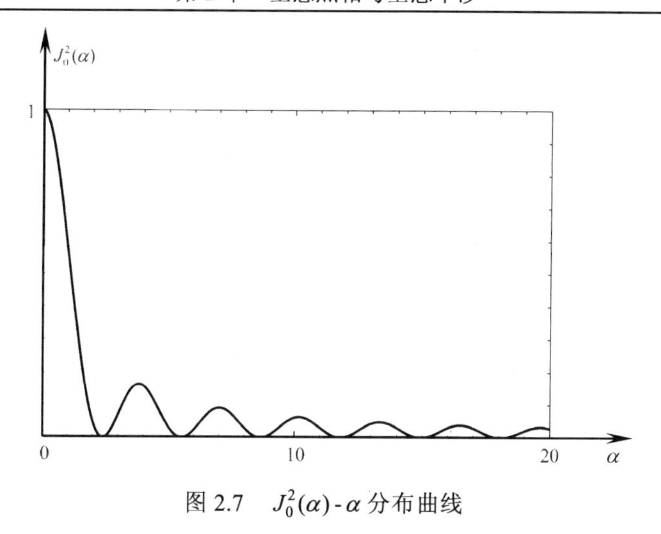

图 2.7　$J_0^2(\alpha)$-α 分布曲线

2.2.5　实时时间平均全息干涉

实时时间平均全息干涉是在物体静止状态下记录全息图，经显影和定影后，全息图精确复位进行再现，让物体发生振动，此时再现物体光波与振动物体的物体光波发生干涉而产生干涉条纹。

设物体在静止状态下的物体光波复振幅为 $O = a_{\text{o}}\exp(\text{i}\varphi_{\text{o}})$，参考光波复振幅为 $R = a_{\text{r}}\exp(\text{i}\varphi_{\text{r}})$。因此，物体在静止状态下，照相底片记录的强度可表示为

$$I = (O+R)(O+R)^* = (a_{\text{o}}^2 + a_{\text{r}}^2) + a_{\text{o}}a_{\text{r}}\exp[\text{i}(\varphi_{\text{o}} - \varphi_{\text{r}})] + a_{\text{o}}a_{\text{r}}\exp[-\text{i}(\varphi_{\text{o}} - \varphi_{\text{r}})] \tag{2.38}$$

设曝光时间为 T，假设振幅透射率与曝光量呈线性关系，取比例常数为 β，则照相底片经显影和定影后，振幅透射率为

$$t = \beta IT = \beta T(a_{\text{o}}^2 + a_{\text{r}}^2) + \beta Ta_{\text{o}}a_{\text{r}}\exp[\text{i}(\varphi_{\text{o}} - \varphi_{\text{r}})] + \beta Ta_{\text{o}}a_{\text{r}}\exp[-\text{i}(\varphi_{\text{o}} - \varphi_{\text{r}})] \tag{2.39}$$

把上述单曝光全息图精确放回原记录系统进行再现，同时用物体光波和参考光波同时照射单曝光全息图。设振动物体在任意时刻的物体光波复振幅为 $O' = a_{\text{o}}\exp[\text{i}(\varphi_{\text{o}} + \delta)]$，则此时透过单曝光全息图的光波复振幅为

$$\begin{aligned} A = (O'+R)t = {} & \beta T(a_{\text{o}}^3 + a_{\text{o}}a_{\text{r}}^2)\exp[\text{i}(\varphi_{\text{o}} + \delta)] + \beta Ta_{\text{o}}^2a_{\text{r}}\exp[\text{i}(2\varphi_{\text{o}} - \varphi_{\text{r}} + \delta)] \\ & + \beta Ta_{\text{o}}^2a_{\text{r}}\exp[\text{i}(\varphi_{\text{r}} + \delta)] + \beta T(a_{\text{o}}^2a_{\text{r}} + a_{\text{r}}^3)\exp(\text{i}\varphi_{\text{r}}) \\ & + \beta Ta_{\text{o}}a_{\text{r}}^2\exp(\text{i}\varphi_{\text{o}}) + \beta Ta_{\text{o}}a_{\text{r}}^2\exp[-\text{i}(\varphi_{\text{o}} - 2\varphi_{\text{r}})] \end{aligned} \tag{2.40}$$

式中，等号右边的第一项 $\beta T(a_{\text{o}}^3 + a_{\text{o}}a_{\text{r}}^2)\exp[\text{i}(\varphi_{\text{o}} + \delta)]$ 和第五项 $\beta Ta_{\text{o}}a_{\text{r}}^2\exp(\text{i}\varphi_{\text{o}})$ 与物体光波有关，取出这两项，得

$$A' = \beta T\exp(\text{i}\varphi_{\text{o}})[a_{\text{o}}a_{\text{r}}^2 + (a_{\text{o}}^3 + a_{\text{o}}a_{\text{r}}^2)\exp(\text{i}\delta)] \tag{2.41}$$

通常 $I_{\text{o}} \ll I_{\text{r}}$，即 $a_{\text{o}}^2 \ll a_{\text{r}}^2$，因此式(2.41)可简化为

$$A' = \beta Ta_{\text{o}}a_{\text{r}}^2\exp(\text{i}\varphi_{\text{o}})[1 + \exp(\text{i}\delta)] \tag{2.42}$$

因此，任意时刻的强度分布为

$$I' = A'A'^* = 2(\beta T a_o a_r^2)^2 (1 + \cos\delta) \tag{2.43}$$

把 $\delta = k(\boldsymbol{e} - \boldsymbol{e}_0) \cdot \boldsymbol{A} \sin \omega t$ 代入式(2.43)，得

$$I' = 2(\beta T a_o a_r^2)^2 \{1 + \cos[k(\boldsymbol{e} - \boldsymbol{e}_0) \cdot \boldsymbol{A} \sin \omega t]\} \tag{2.44}$$

当接收上述强度分布时，所得结果是上述瞬时强度分布的时间平均值。设接收时间为 τ，且 $\tau \gg 2\pi/\omega$，则接收到的强度分布可表示为

$$I_\tau = \frac{1}{\tau}\int_0^\tau I' \mathrm{d}t = 2(\beta T a_o a_r^2)^2 \left\{1 + \frac{1}{\tau}\int_0^\tau \cos[k(\boldsymbol{e} - \boldsymbol{e}_0) \cdot \boldsymbol{A} \sin \omega t]\mathrm{d}t\right\} = 2(\beta T a_o a_r^2)^2 \{1 + J_0[k(\boldsymbol{e} - \boldsymbol{e}_0) \cdot \boldsymbol{A}]\} \tag{2.45}$$

式 (2.45) 表明，实时时间平均法的强度分布与 $\{1 + J_0[k(\boldsymbol{e} - \boldsymbol{e}_0) \cdot \boldsymbol{A}]\}$ 有关。在 $\{1 + J_0[k(\boldsymbol{e} - \boldsymbol{e}_0) \cdot \boldsymbol{A}]\}$ 取极大值处将形成亮纹，当取最大值时，振幅 $A = 0$，即为振动节线。在 $\{1 + J_0[k(\boldsymbol{e} - \boldsymbol{e}_0) \cdot \boldsymbol{A}]\}$ 取极小值处将形成暗纹。因此，通过确定条纹级数即可求得振动物体各点的振幅大小。$[1 + J_0(\alpha)]$-α 的归一化分布曲线如图 2.8 所示。

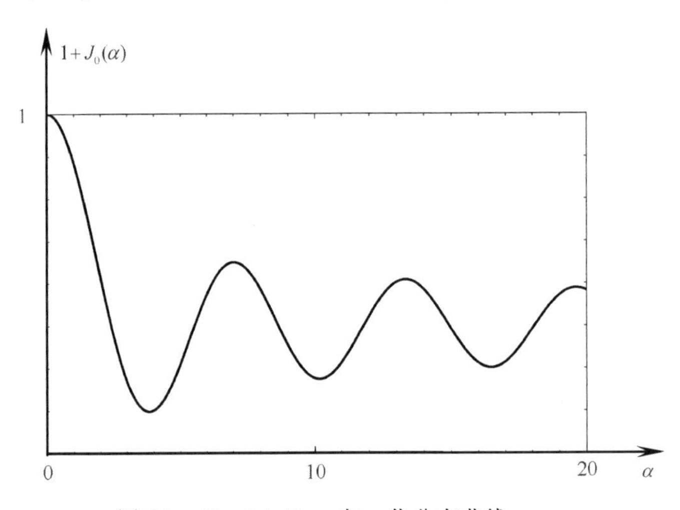

图 2.8 $[1 + J_0(\alpha)]$-α 归一化分布曲线

2.2.6 频闪全息干涉

频闪全息干涉采用与振动物体同步的闪光照明，把振动物体一个周期内的两个瞬时振动状态记录于同一张照相底片上。照相底片经过显影和定影处理后，再放回原记录系统进行再现，则对应于振动物体两个瞬时振动状态的物体光波，因相位不同而发生干涉并形成干涉条纹，通过对干涉条纹进行分析，即可实现物体的振动分析。

设振动物体在任意两个瞬时的物体光波复振幅分别为 $O_1 = a_o \exp[\mathrm{i}(\varphi_o + \delta_1)]$ 和 $O_2 = a_o \exp[\mathrm{i}(\varphi_o + \delta_2)]$，其中，$a_o$ 为物体光波振幅；$(\varphi_o + \delta_1)$ 和 $(\varphi_o + \delta_2)$ 分别为振动物体在两个瞬时的物体光波相位；δ_1 和 δ_2 分别为在两个瞬时因物体振动引起的物体光波的相位变化。设参考光波复振幅为 $R = a_r \exp(\mathrm{i}\varphi_r)$，那么对应于振动物体两个瞬时振动状态，照相底片记录的强度分别为

$$I_1=(O_1+R)\cdot(O_1+R)^*=(a_o^2+a_r^2)+a_o a_r\exp[\mathrm{i}(\varphi_o+\delta_1-\varphi_r)]+a_o a_r\exp[-\mathrm{i}(\varphi_o+\delta_1-\varphi_r)]$$
$$I_2=(O_2+R)\cdot(O_2+R)^*=(a_o^2+a_r^2)+a_o a_r\exp[\mathrm{i}(\varphi_o+\delta_2-\varphi_r)]+a_o a_r\exp[-\mathrm{i}(\varphi_o+\delta_2-\varphi_r)] \quad (2.46)$$

设对应于振动物体两个瞬时振动状态照相底片的单次曝光时间分别为 τ_1 和 τ_2，且振动物体在两个瞬时振动状态的曝光次数分别为 N_1 和 N_2，则照相底片记录到的曝光量可表示为

$$\begin{aligned}E&=I_1N_1\tau_1+I_2N_2\tau_2\\&=(a_o^2+a_r^2)(N_1\tau_1+N_2\tau_2)+a_o a_r\exp[\mathrm{i}(\varphi_o-\varphi_r)][N_1\tau_1\exp(\mathrm{i}\delta_1)+N_2\tau_2\exp(\mathrm{i}\delta_2)]\\&\quad+a_o a_r\exp[-\mathrm{i}(\varphi_o-\varphi_r)][N_1\tau_1\exp(-\mathrm{i}\delta_1)+N_2\tau_2\exp(-\mathrm{i}\delta_2)]\end{aligned} \quad (2.47)$$

照相底片经显影和定影后，设振幅透射率与曝光量呈线性关系，取比例常数为 β，则全息图的振幅透射率为

$$\begin{aligned}t=\beta E=&\beta(a_o^2+a_r^2)(N_1\tau_1+N_2\tau_2)+\beta a_o a_r\exp[\mathrm{i}(\varphi_o-\varphi_r)][N_1\tau_1\exp(\mathrm{i}\delta_1)+N_2\tau_2\exp(\mathrm{i}\delta_2)]\\&+\beta a_o a_r\exp[-\mathrm{i}(\varphi_o-\varphi_r)][N_1\tau_1\exp(-\mathrm{i}\delta_1)+N_2\tau_2\exp(-\mathrm{i}\delta_2)]\end{aligned} \quad (2.48)$$

用参考光波 $R=a_r\exp(\mathrm{i}\varphi_r)$ 照射经显影和定影后的照相底片，则透过全息图的光波复振幅表示为

$$\begin{aligned}A=Rt=&\beta(a_o^2a_r+a_r^3)(N_1\tau_1+N_2\tau_2)\exp(\mathrm{i}\varphi_r)\\&+\beta a_o a_r^2\exp(\mathrm{i}\varphi_o)[N_1\tau_1\exp(\mathrm{i}\delta_1)+N_2\tau_2\exp(\mathrm{i}\delta_2)]\\&+\beta a_o a_r^2\exp[-\mathrm{i}(\varphi_o-2\varphi_r)][N_1\tau_1\exp(-\mathrm{i}\delta_1)+N_2\tau_2\exp(-\mathrm{i}\delta_2)]\end{aligned} \quad (2.49)$$

式中，等号右边的第一项是透过全息图后沿参考光波方向的 0 级衍射光波；第二项是透过全息图后沿物体光波方向的 1 级衍射光波；第三项是物体共轭光波。

仅考虑含有 $\exp(\mathrm{i}\varphi_o)$ 的第二项，则透过全息图的复振幅为

$$A'=\beta a_o a_r^2\exp(\mathrm{i}\varphi_o)[N_1\tau_1\exp(\mathrm{i}\delta_1)+N_2\tau_2\exp(\mathrm{i}\delta_2)] \quad (2.50)$$

相应的强度分布为

$$I'=A'A'^*=(\beta a_o a_r^2)^2[(N_1\tau_1)^2+(N_2\tau_2)^2+2N_1\tau_1N_2\tau_2\cos(\delta_2-\delta_1)] \quad (2.51)$$

设对应于振动物体两个瞬时振动状态的总曝光时间相等，即 $N_1\tau_1=N_2\tau_2=T$，则式(2.51)简化为

$$I'=2(\beta Ta_o a_r^2)^2[1+\cos(\delta_2-\delta_1)] \quad (2.52)$$

这是一般性公式，下面讨论两种特殊情况。

(1) 假设对应于振动物体的两个瞬时振动状态分别为平衡位置和振幅位置，即 $\delta_1=0$ 和 $\delta_2=k(\boldsymbol{e}-\boldsymbol{e}_0)\cdot\boldsymbol{A}$，则式(2.52)可表示为

$$I'=2(\beta Ta_o a_r^2)^2\{1+\cos[k(\boldsymbol{e}-\boldsymbol{e}_0)\cdot\boldsymbol{A}]\} \quad (2.53)$$

式(2.53)表明，当满足条件

$$(\boldsymbol{e}-\boldsymbol{e}_0)\cdot\boldsymbol{A}=n\lambda \quad (n=0,\pm1,\pm2,\cdots) \quad (2.54)$$

时，将形成亮纹；当满足条件

$$(\boldsymbol{e}-\boldsymbol{e}_0)\cdot\boldsymbol{A}=\left(n+\frac{1}{2}\right)\lambda \quad (n=0,\pm1,\pm2,\cdots) \quad (2.55)$$

时，将形成暗纹。

(2) 假设对应于振动物体的两个瞬时振动状态分别为相位相反的 2 个振幅位置，即 $\delta_1 = -k(\boldsymbol{e}-\boldsymbol{e}_0)\cdot\boldsymbol{A}$ 和 $\delta_2 = k(\boldsymbol{e}-\boldsymbol{e}_0)\cdot\boldsymbol{A}$，则式(2.52)可表示为

$$I' = 2(\beta T a_o a_r^2)^2\{1+\cos[2k(\boldsymbol{e}-\boldsymbol{e}_0)\cdot\boldsymbol{A}]\} \tag{2.56}$$

因此，当满足条件

$$(\boldsymbol{e}-\boldsymbol{e}_0)\cdot\boldsymbol{A} = \frac{1}{2}n\lambda \quad (n=0,\pm1,\pm2,\cdots) \tag{2.57}$$

时，将形成亮纹；当满足条件

$$(\boldsymbol{e}-\boldsymbol{e}_0)\cdot\boldsymbol{A} = \frac{1}{2}\left(n+\frac{1}{2}\right)\lambda \quad (n=0,\pm1,\pm2,\cdots) \tag{2.58}$$

时，将形成暗纹。

第 3 章　散斑照相与散斑干涉

当激光照射表面粗糙的物体时，物面就会散射无数相干子波，这些散射子波在物体周围空间相互干涉而形成的无数随机分布的亮点和暗点，称为散斑。

散斑现象早就被人们所发现，但一直未能引起重视。激光诞生后全息技术得到了快速发展，但伴随全息而存在的散斑极大地影响了全息质量，直到此时散斑效应才引起了人们的广泛兴趣。不过当时散斑是被作为全息噪声来进行研究的，从事散斑研究的目的是消除或抑制全息中出现的散斑噪声。但随着对散斑研究的深入，人们发现散斑能够用于变形测量和振动分析。

当激光照射的光学粗糙物面发生位移或变形时，物面周围空间所形成的散斑分布将按一定的规律发生运动或变化。因此通过分析散斑的运动和变化即可测量物体的位移和变形。

3.1　散 斑 照 相

散斑照相由粗糙物体的随机散射子波之间的干涉而形成。散斑照相包括双曝光散斑照相、时间平均散斑照相和频闪散斑照相等。

3.1.1　双曝光散斑照相

双曝光散斑照相要求对两个瞬时散斑场进行双曝光记录，即物体变形前后散斑场记录在同一张散斑图上。对显影和定影后的散斑图进行滤波(逐点滤波或全场滤波)，可以获得记录在散斑图上的散斑位移，再通过物点和像点之间的位移换算关系得到物体的位移或变形。

1. 散斑图记录

双曝光散斑照相记录系统如图 3.1 所示。用一束激光(也可采用白光或部分相干光)照射物面，在像面记录散斑图。通过两次系列曝光把对应于物体位移或变形前后的两个状态记录于同一张散斑图上。

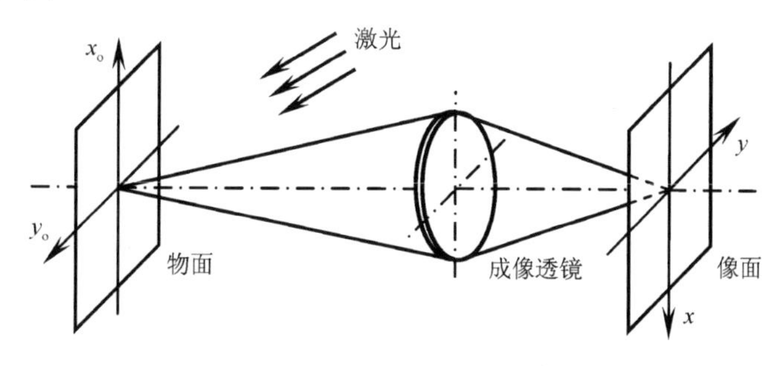

图 3.1　记录系统

设物体变形前后像面强度分布分别由 $I_1(x,y)$ 和 $I_2(x,y)$ 表示，则

$$I_2(x,y)=I_1(x-u,y-v) \tag{3.1}$$

式中，$u=u(x,y)$ 和 $v=v(x,y)$ 为散斑图上点 (x,y) 处分别沿 x 和 y 方向的位移分量。

设物体变形前后曝光时间均为 T，则散斑图的曝光量可表示为

$$E(x,y)=T[I_1(x,y)+I_2(x,y)] \tag{3.2}$$

经显影和定影后，在一定曝光量范围内，散斑图的振幅透射率与曝光量呈线性关系，若取比例常数为 β，那么散斑图的振幅透射率为

$$t(x,y)=\beta E(x,y)=\beta T[I_1(x,y)+I_2(x,y)] \tag{3.3}$$

2. 散斑图滤波

把散斑图放入如图 3.2 所示的滤波系统进行处理，用单位振幅的平行激光照射散斑图，则在傅里叶变换面上的频谱分布为

$$\mathrm{FT}\{t(x,y)\}=\beta T[\mathrm{FT}\{I_1(x,y)\}+\mathrm{FT}\{I_2(x,y)\}] \tag{3.4}$$

式中，$\mathrm{FT}\{\cdot\}$ 表示傅里叶变换。

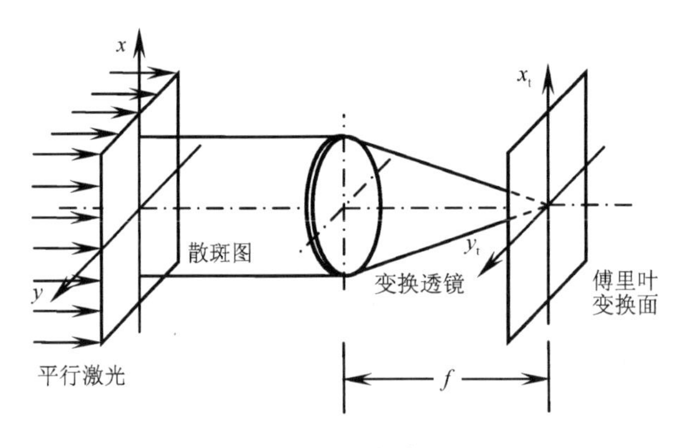

图 3.2 滤波系统

利用傅里叶变换的平移性质，可得

$$\mathrm{FT}\{I_2(x,y)\}=\mathrm{FT}\{I_1(x-u,y-v)\}=\mathrm{FT}\{I_1(x,y)\}\exp[-\mathrm{i}2\pi(uf_x+vf_y)] \tag{3.5}$$

式中，$(f_x,f_y)=(x_t,y_t)/(\lambda f)$ 为傅里叶变换面上分别沿 x 和 y 方向的频率坐标，其中，λ 为激光波长，f 为变换透镜焦距；(x_t,y_t) 为傅里叶变换面上分别沿 x 和 y 方向的坐标。

利用式(3.5)，则式(3.4)可表示为

$$\mathrm{FT}\{t(x,y)\}=\beta T\mathrm{FT}\{I_1(x,y)\}\{1+\exp[-\mathrm{i}2\pi(uf_x+vf_y)]\} \tag{3.6}$$

因此，散斑图经过傅里叶变换后，在傅里叶变换平面上的衍射晕强度分布为

$$I(f_x,f_y)=|\mathrm{FT}\{t(x,y)\}|^2=2\beta^2T^2\,|\mathrm{FT}\{I_1(x,y)\}|^2\,\{1+\cos[2\pi(uf_x+vf_y)]\} \tag{3.7}$$

由式(3.8)可见，在双曝光衍射晕中将出现余弦干涉条纹。

如果散斑图上各点的位移矢量相同，则在傅里叶变换平面上将出现干涉条纹。然而，通常散斑图上各点的位移互不相同，此时在傅里叶变换平面上出现的是各种间隔和各种方向的干涉条纹的叠加，因而一般不能直接观察到干涉条纹，但通过逐点滤波或全场滤波即可显现干涉条纹。

3. 逐点滤波

用细激光束照射散斑图上的点 P，如图 3.3 所示。当照射区域较小时，在观察面上可以直接呈现干涉条纹，即杨氏条纹。

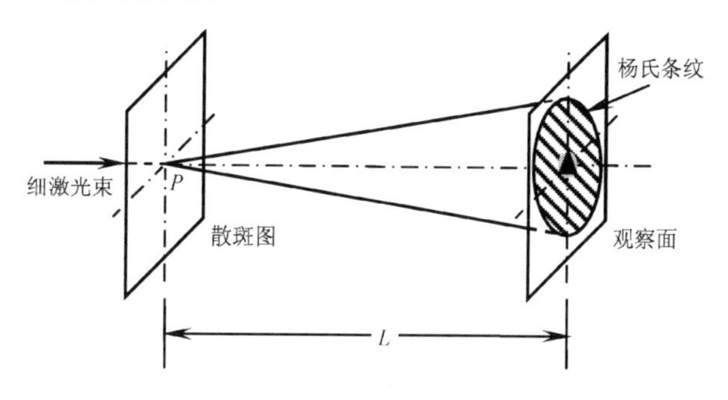

图 3.3　逐点滤波

散斑位移方向垂直于杨氏条纹，散斑位移大小反比于杨氏条纹间距，则散斑图上点 P 处的位移大小可表示为

$$d = \frac{\lambda L}{\Delta} \tag{3.8}$$

式中，Δ 为杨氏条纹间距；L 为观察面到散斑图的距离。

4. 全场滤波

全场滤波系统如图 3.4 所示。在傅里叶变换面放置开有滤波孔的不透光屏，通过滤波孔进行观察，就能看到干涉条纹。当滤波孔沿径向移动时，干涉条纹的疏密在连续变化，滤波孔离光轴越远条纹越密；当滤波孔沿周向移动时，干涉条纹的方向在连续变化。

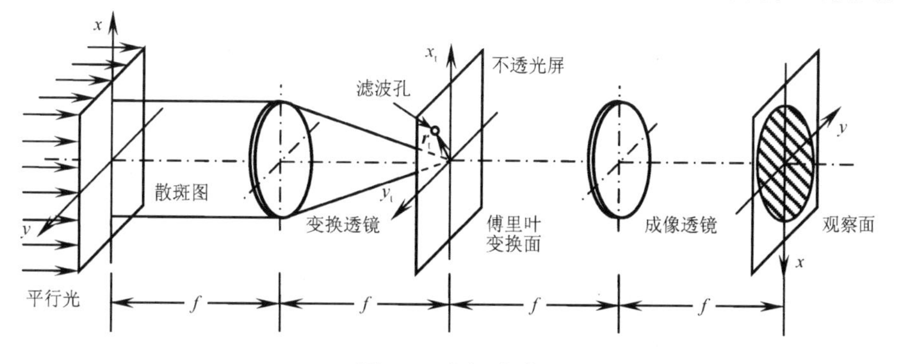

图 3.4　全场滤波

全场滤波所得的干涉条纹表示散斑图上各点沿滤波孔方向的位移等值线。当滤波孔处于位置 $(x_t,0)$ 时，由式(3.7)可知，亮纹出现的条件为

$$u = \frac{m}{f_x} = \frac{m\lambda f}{x_t} \quad (m = 0, \pm 1, \pm 2, \cdots) \tag{3.9}$$

同理，当滤波孔处于位置$(0, y_t)$，满足条件

$$v = \frac{n}{f_y} = \frac{n\lambda f}{y_t} \quad (n = 0, \pm 1, \pm 2, \cdots) \tag{3.10}$$

时，将产生亮纹。

3.1.2 时间平均散斑照相

时间平均散斑照相是指对稳态振动物体，采用比振动周期长得多的时间进行连续曝光，记录散斑图。时间平均法把振动物体的所有振动状态都记录在同一张散斑图上，散斑图经过显影和定影处理后进行滤波将得到表征物体振幅分布的干涉条纹。

1. 散斑图记录

设在任意时刻的像面强度分布为$I(x-u, y-v)$，其中$u = u(x, y; t)$和$v = v(x, y; t)$为散斑图上点(x, y)处在时刻t分别沿x和y方向的位移分量。设曝光时间为T，则散斑图的曝光量可表示为

$$E(x, y) = \int_0^T I(x-u, y-v)\mathrm{d}t \tag{3.11}$$

经显影和定影后，在一定曝光量范围内，散斑图的振幅透射率与曝光量呈线性关系，若取比例常数为β，那么散斑图的振幅透射率为

$$t(x, y) = \beta E(x, y) = \beta\int_0^T I(x-u, y-v)\mathrm{d}t \tag{3.12}$$

2. 散斑图滤波

把散斑图放入全场滤波系统中进行分析，用单位振幅平行激光照射散斑图，则在傅里叶变换面上的频谱分布为

$$\mathrm{FT}\{t(x, y)\} = \beta\int_0^T \mathrm{FT}\{I(x-u, y-v)\}\mathrm{d}t \tag{3.13}$$

利用傅里叶变换的平移性质，可得

$$\mathrm{FT}\{I(x-u, y-v)\} = \mathrm{FT}\{I(x, y)\}\exp[-\mathrm{i}2\pi(uf_x + vf_y)] \tag{3.14}$$

利用式(3.14)，则式(3.13)可表示为

$$\mathrm{FT}\{t(x, y)\} = \beta\mathrm{FT}\{I(x, y)\}\int_0^T \exp[-\mathrm{i}2\pi(uf_x + vf_y)]\mathrm{d}t \tag{3.15}$$

因此，散斑图经过傅里叶变换后，在傅里叶变换平面上的衍射晕强度分布为

$$I(f_x, f_y) = |\mathrm{FT}\{t(x, y)\}|^2 = \beta^2 |\mathrm{FT}\{I(x, y)\}|^2 \left|\int_0^T \exp[-\mathrm{i}2\pi(uf_x + vf_y)]\mathrm{d}t\right|^2 \tag{3.16}$$

设物体发生简谐振动，则散斑图上各点在时刻 t 分别沿 x 和 y 方向的位移分量可表示为

$$u = A_x \sin \omega t, \quad v = A_y \sin \omega t \tag{3.17}$$

式中，$A_x = A_x(x,y)$ 和 $A_y = A_y(x,y)$ 为散斑图上点 (x,y) 处分别沿 x 和 y 方向的振幅分量；ω 为圆频率。把式(3.17)代入式(3.16)，得

$$I(f_x, f_y) = \beta^2 \,|\, \mathrm{FT}\{I(x,y)\}\,|^2 \left| \int_0^T \exp[-\mathrm{i}2\pi(A_x f_x + A_y f_y)\sin \omega t]\mathrm{d}t \right|^2 \tag{3.18}$$

当 $T >> 2\pi/\omega$ 时，式(3.18)可表示为

$$I(f_x, f_y) = \beta^2 T^2 \,|\, \mathrm{FT}\{I(x,y)\}\,|^2 \, J_0^2[2\pi(A_x f_x + A_y f_y)] \tag{3.19}$$

式中，J_0 为第一类零阶贝塞尔函数。由式(3.19)可见，在衍射晕中将出现干涉条纹。通常，散斑图上各点的振幅矢量互不相同，此时傅里叶变换平面上出现的是各种间隔和各种方向的干涉条纹的叠加，因而在傅里叶变换平面上一般不能直接观察到干涉条纹，但通过全场滤波即可提取这些干涉条纹。

式(3.19)表明，在 $J_0^2[2\pi(A_x f_x + A_y f_y)]$ 取极大值处将形成亮纹。特别是当 $J_0^2[2\pi(A_x f_x + A_y f_y)]$ 取最大值时振幅分量 $A_x = A_y = 0$，即条纹最亮处为振动节线。在 $J_0^2[2\pi(A_x f_x + A_y f_y)]$ 取极小值处将形成暗纹，即当满足条件

$$2\pi(A_x f_x + A_y f_y) = \alpha \quad (\alpha = 2.41,\ 5.52,\ 8.65,\ 11.79,\ 14.98,\ \cdots) \tag{3.20}$$

时将产生暗纹。$J_0^2(\alpha)$-α 的分布曲线如图 3.5 所示。

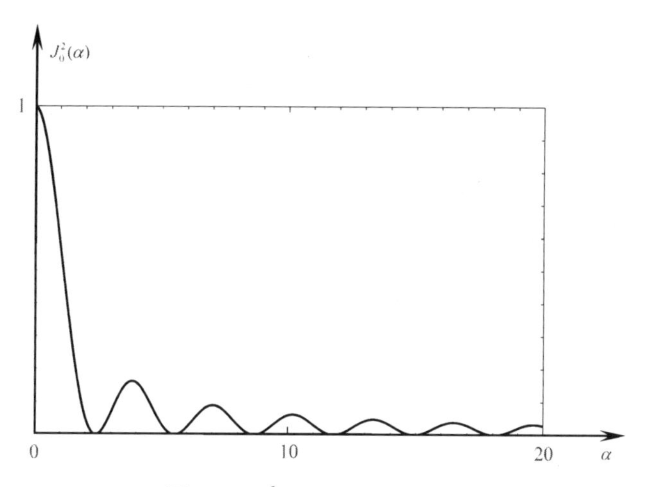

图 3.5　$J_0^2(\alpha)$-α 的分布曲线

当滤波孔处于位置 $(x_t, 0)$ 和 $(0, y_t)$ 时，由式(3.19)知，出现暗纹的条件分别为

$$\begin{aligned} A_x &= \frac{\alpha_x}{2\pi f_x} = \frac{\alpha_x \lambda f}{2\pi x_t} \quad (\alpha_x = 2.41,\ 5.52,\ 8.65,\ 11.79,\ 14.98,\ \cdots) \\ A_y &= \frac{\alpha_y}{2\pi f_y} = \frac{\alpha_y \lambda f}{2\pi y_t} \quad (\alpha_y = 2.41,\ 5.52,\ 8.65,\ 11.79,\ 14.98,\ \cdots) \end{aligned} \tag{3.21}$$

3.1.3 频闪散斑照相

频闪散斑照相把动态变形物体的两个瞬时状态记录于同一张散斑图上。散斑图经过显影和定影后进行滤波得到对应于两个瞬时状态的干涉条纹，通过对干涉条纹进行分析，可实现动态测量。

1. 散斑图记录

设动态变形物体在两个瞬时状态的像面强度分布分别为 $I(x-u_1, y-v_1)$ 和 $I(x-u_2, y-v_2)$，两个瞬时状态的曝光时间均为 τ，则散斑图的曝光量可表示为

$$E(x,y)=\tau[I(x-u_1,y-v_1)+I(x-u_2,y-v_2)] \tag{3.22}$$

式中，$u_1=u_1(x,y;t_1)$ 和 $v_1=v_1(x,y;t_1)$ 为散斑图上点 (x,y) 处在时刻 t_1 分别沿 x 和 y 方向的位移分量；$u_2=u_2(x,y;t_2)$ 和 $v_2=v_2(x,y;t_2)$ 为散斑图上点 (x,y) 处在时刻 t_2 分别沿 x 和 y 方向的位移分量。

经显影和定影后，在一定曝光量范围内，散斑图的振幅透射率与曝光量呈线性关系，若取比例常数为 β，那么散斑图的振幅透射率为

$$t(x,y)=\beta E(x,y)=\beta\tau[I(x-u_1,y-v_1)+I(x-u_2,y-v_2)] \tag{3.23}$$

2. 散斑图滤波

把散斑图放入滤波系统进行分析，用单位振幅平行激光照射散斑图，则在傅里叶变换面上的频谱分布为

$$\mathrm{FT}\{t(x,y)\}=\beta\tau[\mathrm{FT}\{I(x-u_1,y-v_1)\}+\mathrm{FT}\{I(x-u_2,y-v_2)\}] \tag{3.24}$$

利用傅里叶变换的平移性质，则

$$\mathrm{FT}\{t(x,y)\}=\beta\tau[\mathrm{FT}\{I(x-u_1,y-v_1)\}+\mathrm{FT}\{I(x-u_2,y-v_2)\}] \tag{3.25}$$

利用式(3.25)，则式(3.24)可表示为

$$\begin{aligned}\mathrm{FT}\{I(x-u_1,y-v_1)\}&=\mathrm{FT}\{I(x,y)\}\exp[-\mathrm{i}2\pi(u_1f_x+v_1f_y)]\\ \mathrm{FT}\{I(x-u_2,y-v_2)\}&=\mathrm{FT}\{I(x,y)\}\exp[-\mathrm{i}2\pi(u_2f_x+v_2f_y)]\end{aligned} \tag{3.26}$$

因此，散斑图经过傅里叶变换，在傅里叶变换平面上的衍射晕强度分布为

$$I(f_x,f_y)=|\mathrm{FT}\{t(x,y)\}|^2=2\beta^2\tau^2\,|\mathrm{FT}\{I(x,y)\}|^2\,\{1+\cos[2\pi(uf_x+vf_y)]\} \tag{3.27}$$

式中，$u=u_2(x,y;t_2)-u_1(x,y;t_1)$ 和 $v=v_2(x,y;t_2)-v_1(x,y;t_1)$ 为散斑图上点 (x,y) 处在两个瞬时 t_1 和 t_2 分别沿 x 和 y 方向的相对位移分量。由式（3.27）可见，通过逐点滤波或全场滤波可以提取这些干涉条纹。采用全场滤波，当滤波孔处于位置 $(x_t,0)$ 和 $(0,y_t)$ 时，亮纹出现的条件分别为

$$u=\frac{m}{f_x}=\frac{m\lambda f}{x_t}\quad(m=0,\ \pm1,\ \pm2,\ \cdots),\quad v=\frac{n}{f_y}=\frac{n\lambda f}{y_t}\quad(n=0,\ \pm1,\ \pm2,\ \cdots) \tag{3.28}$$

3.2 散 斑 干 涉

散斑干涉是由表面粗糙物体的随机散射子波与另一参考光波之间的干涉而形成的。

3.2.1 面内位移测量

图 3.6 所示为面内位移测量系统。用垂直于 y_o 轴平面内的两束激光对称照射物面，物体变形前在照相底片上进行第一次曝光记录，物体变形后在同样的照相底片上进行第二次曝光记录。对散斑图进行滤波，将产生沿 x 方向的面内位移分量等值条纹。

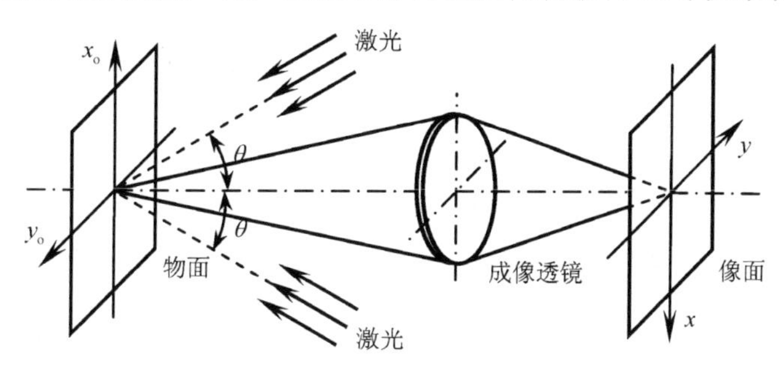

图 3.6　面内位移测量系统

物体变形前像面强度分布可表示为

$$I_1(x,y)=I_{o1}+I_{o2}+2\sqrt{I_{o1}I_{o2}}\cos\varphi \tag{3.29}$$

式中，I_{o1} 和 I_{o2} 分别为对应于两入射光波的强度分布；φ 为两入射光波之间的相位差。

同理，物体变形后像面强度分布为

$$I_2(x,y)=I_{o1}+I_{o2}+2\sqrt{I_{o1}I_{o2}}\cos(\varphi+\delta) \tag{3.30}$$

式中，$\delta=\delta_1-\delta_2$，δ_1 和 δ_2 是由于物体变形而引起的两入射光波的相位变化。根据式(2.8)，δ_1 和 δ_2 可分别表示为

$$\delta_1=k[w_o(1+\cos\theta)+u_o\sin\theta],\quad \delta_2=k[w_o(1+\cos\theta)-u_o\sin\theta] \tag{3.31}$$

式中，k 为激光波数；θ 为两束入射光与光轴之间的夹角；u_o 和 w_o 分别为物面上沿 x 和 z 方向的位移分量。因此，由物体变形引起的两入射光波的相对相位变化为

$$\delta=2ku_o\sin\theta \tag{3.32}$$

设对应物体变形前后的两次曝光时间均为 T，如果振幅透射率与曝光量呈线性关系，取比例常数为 β，那么散斑图的振幅透射率可表示为

$$\begin{aligned}t(x,y)&=\beta T[I_1(x,y)+I_2(x,y)]\\&=2\beta T(I_{o1}+I_{o2})+4\beta T\sqrt{I_{o1}I_{o2}}\cos\left(\varphi+\frac{1}{2}\delta\right)\cos\left(\frac{1}{2}\delta\right)\end{aligned} \tag{3.33}$$

式中，φ 为随机快变化函数；δ 为慢变化函数。含有 φ 的余弦函数为高频成分，仅含有 δ 的余弦函数为低频成分。当满足条件

$$\cos\left(\frac{1}{2}\delta\right)=0 \tag{3.34}$$

即

$$\delta=(2n+1)\pi \quad (n=0,\pm1,\pm2,\cdots) \tag{3.35}$$

时，将得到暗纹。把式(3.32)代入式(3.35)，得

$$u_o=\frac{(2n+1)\pi}{2k\sin\theta}=\frac{(2n+1)\lambda}{4\sin\theta} \quad (n=0,\pm1,\pm2,\cdots) \tag{3.36}$$

由于存在背景强度 $2\beta T(I_{o1}+I_{o2})$，因此直接从上述散斑图看不到干涉条纹。对散斑图进行全场滤波，通过对滤掉低频分量的衍射晕进行观察，则在观察面上可显现面内位移分量等值条纹。

3.2.2　离面位移测量

离面位移测量系统如图 3.7 所示。像面强度因参考光波同物面散斑场的干涉而产生。在物体变形前后分别进行曝光，所得到的散斑图在进行滤波时将得到离面位移等值条纹。

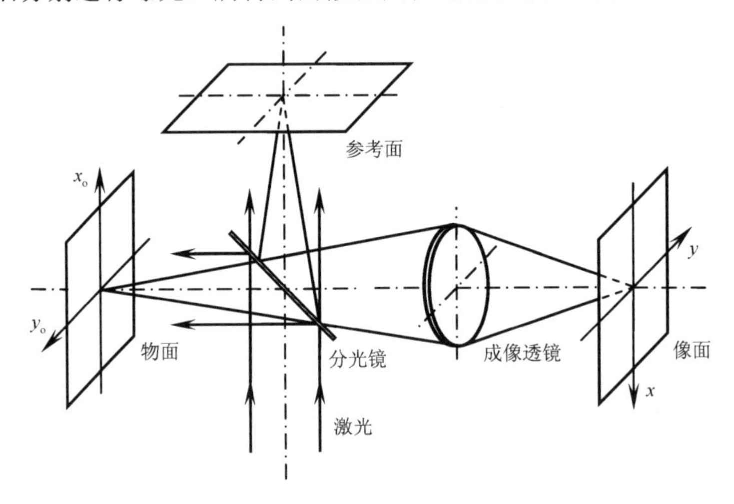

图 3.7　离面位移测量系统

对应第一次曝光的强度分布为

$$I_1(x,y)=I_o+I_r+2\sqrt{I_oI_r}\cos\varphi \tag{3.37}$$

式中，I_o 和 I_r 分别为对应于物体光波和参考光波的强度分布；φ 为物体光波和参考光波的相位差。

对应第二次曝光的强度分布为

$$I_2(x,y)=I_o+I_r+2\sqrt{I_oI_r}\cos(\varphi+\delta) \tag{3.38}$$

式中，δ 为因物体变形引起的物体光波和参考光波的相对相位变化。当物体光波垂直照射和垂直接收时，相位变化 δ 表示为

$$\delta=2kw_o \tag{3.39}$$

式中，w_o 为离面位移。

经过两次曝光，散斑图的强度分布可表示为

$$I(x,y)=I_1(x,y)+I_2(x,y)=2(I_o+I_r)+4\sqrt{I_oI_r}\cos\left(\varphi+\frac{\delta}{2}\right)\cos\frac{\delta}{2} \tag{3.40}$$

式中，$\cos\left(\varphi+\frac{\delta}{2}\right)$ 为高频成分；$\cos\frac{\delta}{2}$ 为低频成分。当满足条件

$$\delta=(2n+1)\pi \quad (n=0,\pm1,\pm2,\cdots) \tag{3.41}$$

时，将形成暗纹。把式(3.39)代入式(3.41)，得

$$w_o=\frac{(2n+1)\pi}{2k}=\frac{(2n+1)\lambda}{4} \quad (n=0,\pm1,\pm2,\cdots) \tag{3.42}$$

由于存在背景强度，因此直接从上述散斑图看不到干涉条纹。滤掉低频分量后，可观察到离面位移分量等值条纹。

3.3　散斑剪切干涉

散斑剪切干涉是由表面粗糙物体相互错位的随机散射子波之间的干涉而形成的，因此散斑剪切干涉有时也称为散斑错位干涉。散斑剪切干涉可以测量离面位移导数(即斜率)。

离面位移导数测量系统如图 3.8 所示。双孔屏对称放置于成像透镜前(双孔沿 x 轴)，在一个孔上放置剪切方向沿 x 轴的剪切镜，用一束激光照射物面，则物面上的一点将成像于像面上两点，或者说物面上的邻近两点将成像于像面上的同一点。变形前进行一次曝光，变形后进行第二次曝光。散斑图在进行滤波时将产生沿双孔方向的全场斜率等值条纹。

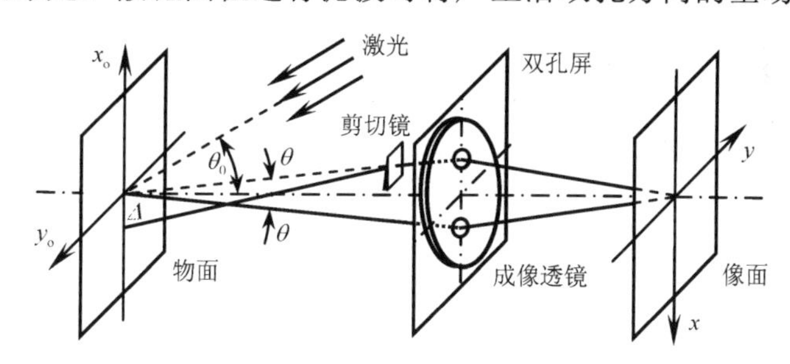

图 3.8　散斑剪切干涉系统

物面剪切量可表示为

$$\Delta=d_o(\mu-1)\alpha \tag{3.43}$$

式中，d_o 是物面到剪切镜的距离；μ 和 α 分别为剪切镜的折射率和楔角。

设通过两孔来自物面的光波强度分布分别为 I_{o1} 和 I_{o2}，那么物体变形前的像面强度分布为

$$I_1(x,y)=I_{o1}+I_{o2}+2\sqrt{I_{o1}I_{o2}}\cos(\varphi+\beta) \tag{3.44}$$

式中，φ 为对应于两物点的相对相位；β 为通过两孔的光波因干涉而产生的栅线结构相位。

同理，变形后的像面强度分布为

$$I_2(x,y)=I_{o1}+I_{o2}+2\sqrt{I_{o1}I_{o2}}\cos(\varphi+\delta+\beta) \tag{3.45}$$

式中，$\delta=\delta_1-\delta_2$，δ_1和δ_2分别为因物体变形引起的通过两孔的散射光波的相位变化。根据照射和观察方向，δ可表示为

$$\delta=k\left\{2u_o\sin\theta+\left[(\sin\theta+\sin\theta_0)\frac{\partial u_o}{\partial x}+(\cos\theta+\cos\theta_0)\frac{\partial w_o}{\partial x}\right]\varDelta\right\} \tag{3.46}$$

式中，θ_0和θ分别为照射方向和观察方向与光轴之间的夹角；u_o为沿x方向的面内位移分量；$\partial u_o/\partial x$和$\partial w_o/\partial x$分别为沿x方向的面内位移导数和离面位移导数。利用$\sin\theta<<1$，式(3.46)可简化为

$$\delta=k\left[\sin\theta_0\frac{\partial u_o}{\partial x}+(1+\cos\theta_0)\frac{\partial w_o}{\partial x}\right]\varDelta \tag{3.47}$$

当激光垂直于物面照射，即$\theta_0=0$时，式(3.47)进一步简化为

$$\delta=2k\frac{\partial w_o}{\partial x}\varDelta \tag{3.48}$$

经过两次曝光，像面强度分布为

$$I(x,y)=I_1(x,y)+I_2(x,y)=2(I_{o1}+I_{o2})+2\sqrt{I_{o1}I_{o2}}[\cos(\varphi+\beta)+\cos(\varphi+\delta+\beta)] \tag{3.49}$$

对散斑图进行滤波，让1级衍射晕通过，则观察面上的强度分布可表示为

$$I(x_t,y_t)=2I_{o1}I_{o2}(1+\cos\delta) \tag{3.50}$$

由此可见，当满足条件$\delta=2n\pi$，即

$$\frac{\partial w_o}{\partial x}=\frac{n\pi}{k\varDelta}=\frac{n\lambda}{2\varDelta}\quad(n=0,\pm1,\pm2,\cdots) \tag{3.51}$$

时，将形成亮纹。

第4章　几何云纹与云纹干涉

云纹图是指波纹图，与中国制造的丝绸织物有关。古代华人挤压两层丝绸而在丝绸织物上形成云纹图。欧洲丝绸经销商引进这些丝绸织物后开始使用术语“云纹”。

4.1　几何云纹

当纱窗或光栅等周期性结构相互重叠时，即可在重叠区域形成由亮暗相间条纹构成的云纹图。出现在云纹图中的亮暗条纹称为云纹条纹。

因为云纹条纹对微小变形或转动极其敏感，因此云纹法在许多不同领域具有广泛应用。例如，云纹法可用于检测物体的微小变形，也可用于测量物体的微小转角。几何云纹是利用两块光栅相互重叠时所形成的云纹条纹进行面内和离面位移测量。

几何云纹所用的基本测量元件是振幅光栅，由不透光和透光线条构成。不透光线条称为栅线，相邻栅线之间的间隔称为节距，如图4.1所示。

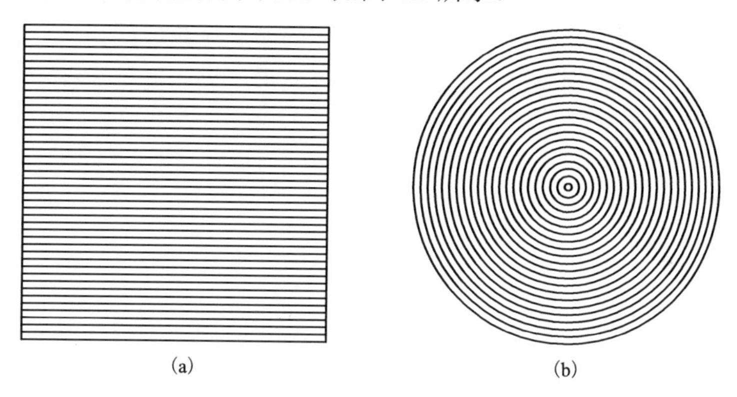

图4.1　光栅

虽然几何云纹中常用光栅为直线栅线(图4.1(a))，然而栅线也可以是曲线的(图4.1(b))。

4.1.1　几何云纹形成

把两块完全相同的光栅重叠起来，如果栅线完全重合，此时重叠区域与一块光栅一样，不会出现任何条纹，如图4.2(a)所示。但当两块光栅有相对转动(图4.2(b))或其中一块光栅的节距增大(图4.2(c))或减小(图4.2(d))时，在重叠区域，一块光栅的不透光部分将遮盖另一块光栅的透光部分，从而形成比栅线宽得多的暗纹，而在两块光栅都透光的部分，将形成亮纹。当栅线有相对转动或节距有相对变化时，云纹条纹会随着栅线相对转动或节距相对变化而发生改变。因此，通过测量云纹条纹即可得到位移分布。

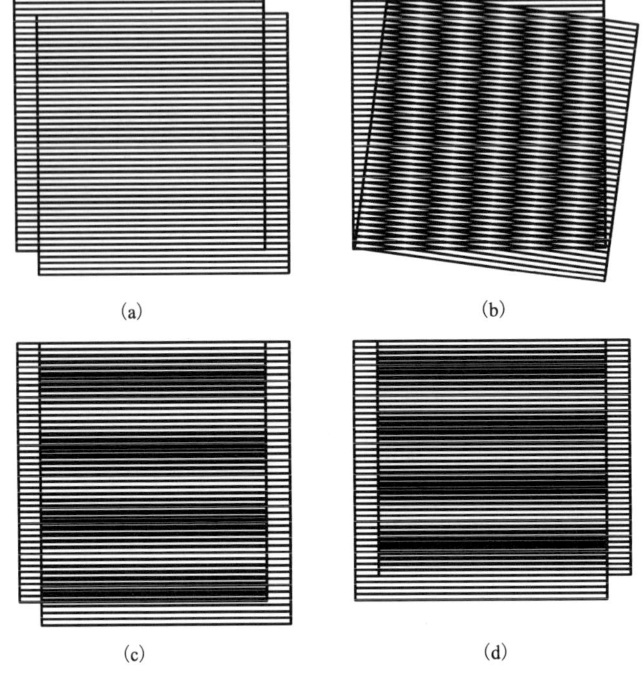

图 4.2　云纹形成

云纹方法通常需要两块光栅，一块固定在物体的测试区域，随物体一起变形，该光栅称为试件栅；另一块与试件栅重叠，不随物体一起变形，该光栅称为参考栅。

4.1.2　几何云纹应变测量

1. 拉压应变测量

1) 平行云纹法

具有相同节距的试件栅和参考栅的栅线垂直于物体的拉伸或压缩方向，并让试件栅和参考栅的栅线完全重合，此时没有云纹条纹产生。物体均匀拉伸或压缩后，试件栅将随物体发生同样的均匀拉伸或压缩变形，此时将有云纹条纹产生，如图 4.3 所示。

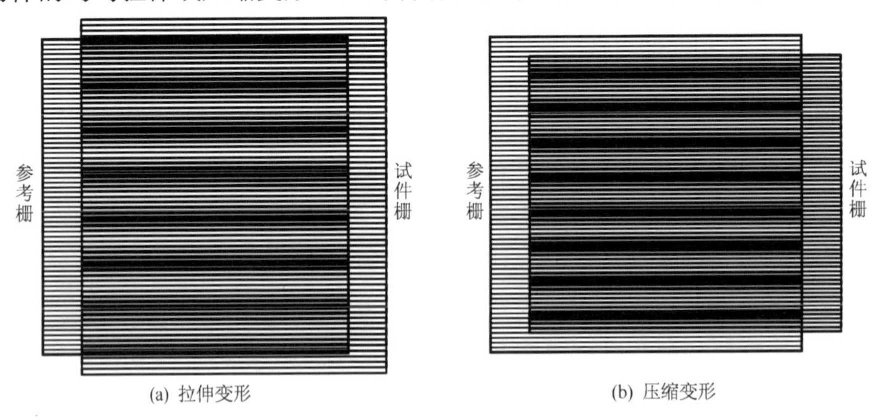

图 4.3　平行云纹

设物体变形前试件栅的节距为 p，如果均匀拉伸或压缩应变为 ε（拉伸时 $\varepsilon>0$，压缩时 $\varepsilon<0$），则试件栅变形后的节距为 $p'=(1+\varepsilon)p$。设相邻云纹条纹间距为 f，则相邻云纹条纹之间含有 $n=f/p$ 条试件栅变形前的栅线（或参考栅的栅线）和 $n'=f/p\mp1$ 条试件栅变形后的栅线（拉伸取“－”号，压缩取“＋”号），因此存在关系：$f=n'p'=(f/p\mp1)(1+\varepsilon)p$，求解得

$$\varepsilon=\frac{\pm p}{f\mp p} \tag{4.1}$$

考虑到 $p \ll f$，由此得

$$\varepsilon=\pm\frac{p}{f} \tag{4.2}$$

式中，“＋”号对应于均匀拉伸；“－”号对应于均匀压缩。节距 p 已知，因此只要测量相邻云纹条纹间距 f，即可求得垂直于栅线方向的均匀拉应变和均匀压应变。

2）转角云纹法

试件栅的栅线垂直于物体的拉伸或压缩方向，参考栅与试件栅的夹角为 θ（θ 逆时针为正），如图 4.4 所示。设参考栅的节距为 q，试件栅变形前的节距为 p，变形后的节距为 p'，则应变为

$$\varepsilon=\frac{p'-p}{p}=\frac{p'}{p}-1=\frac{q}{p}\frac{p'}{q}-1 \tag{4.3}$$

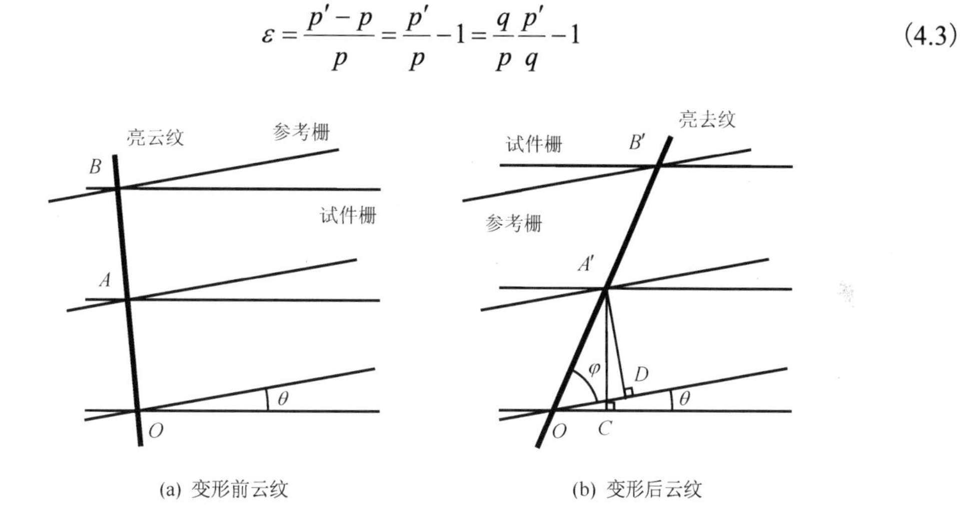

(a) 变形前云纹　　(b) 变形后云纹

图 4.4　转角云纹

变形前亮云纹条纹为 OAB，变形后亮云纹条纹则为 $OA'B'$。设变形后云纹条纹与参考栅的栅线之间的夹角为 φ（φ 逆时针为正），则由图 4.4 可得

$$p'=A'C=OA'\sin(\theta+\varphi),\quad q=A'D=OA'\sin\varphi \tag{4.4}$$

将式(4.4)代入式(4.3)，得

$$\varepsilon=\frac{p'-p}{p}=\frac{p'}{p}-1=\frac{q}{p}\frac{\sin(\theta+\varphi)}{\sin\varphi}-1 \tag{4.5}$$

式中，p、q和θ已知，所以只要测得φ，即可求得应变ε。

如果参考栅和变形前试件栅的节距相等，则式(4.5)可简化为

$$\varepsilon=\frac{\sin(\theta+\varphi)}{\sin\varphi}-1 \tag{4.6}$$

2. 剪切应变测量

剪切应变测量分两步进行，第一步将参考栅和试件栅的栅线平行于x方向放置，设其节距均为p，如图4.5所示。当试件栅发生剪切变形时，试件栅的节距p保持不变，θ_y仅使试件栅的栅线沿x方向移动，θ_x使试件栅的栅线发生转动。因此，只有θ_x才会引起云纹条纹。由图中$BC\perp AC$，得

$$\sin\theta_x=\frac{BC}{AB}=\frac{p}{f_x} \tag{4.7}$$

式中，f_x为云纹条纹在x方向(即栅线方向)的间距。利用$\sin\theta_x\approx\theta_x$，式(4.7)表示为

$$\theta_x=\frac{p}{f_x} \tag{4.8}$$

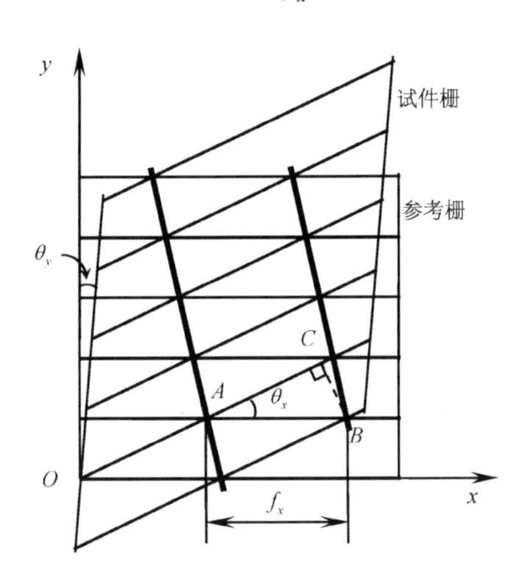

图4.5　切应变测量

第二步将参考栅和试件栅的栅线平行于y方向放置，当试件栅发生剪切变形时，试件栅的节距p保持不变，θ_x仅使试件栅的栅线沿y方向移动而不产生云纹条纹，θ_y使试件栅栅线转动而产生云纹条纹。同理可得

$$\theta_y=\frac{p}{f_y} \tag{4.9}$$

式中，f_y为云纹条纹在y方向(即栅线方向)的间距。综合式(4.8)和式(4.9)，得剪切应变

$$\gamma_{xy}=\theta_x+\theta_y=\frac{p}{f_x}+\frac{p}{f_y} \tag{4.10}$$

3. 平面应变测量

平面应变同时有 ε_x、ε_y 和 γ_{xy}。平面应变测量也分两步进行。第一步将参考栅和试件栅的栅线平行于 x 方向放置，设其节距均为 p，如图 4.6 所示。当物体发生平面变形时，试件栅也随之发生同样的平面变形，设变形后试件栅的节距为 p'。ε_x 仅使试件栅的栅线沿 x 方向伸长或缩短，θ_y 仅使试件栅的栅线沿 x 方向移动，因此 ε_x 和 θ_y 对云纹条纹没有影响。ε_y 使试件栅的节距增大或减小，θ_x 使试件栅的栅线转动，因此 ε_y 和 θ_x 将产生云纹条纹。由图得

$$\sin\theta_x = \frac{BC}{AB} = \frac{p'}{f_x} \tag{4.11}$$

式中，f_x 为云纹条纹在 x 方向(即栅线方向)的间距。利用 $\sin\theta_x \approx \theta_x$ 和 $p' = (1+\varepsilon_y)p \approx p$，式(4.11)简化为

$$\theta_x = \frac{p}{f_x} \tag{4.12}$$

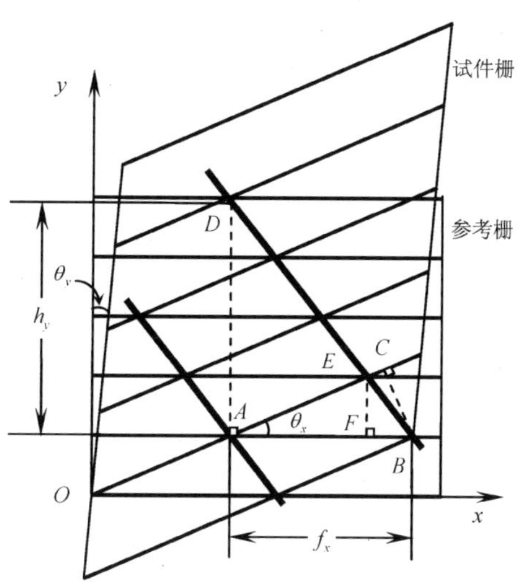

图 4.6　平面应变测量

另外，由图中△*FBE* 与△*ABD* 相似，得

$$FB = FE\cdot\frac{AB}{AD} = p\cdot\frac{f_x}{h_y} \tag{4.13}$$

式中，h_y 为云纹条纹在 y 方向(即垂直栅线方向)的间距。再由图中△*AEF* 与△*ABC* 相似，得

$$FB = AB - AF = AB - EF\cdot\frac{AC}{BC} = AB - EF\cdot\frac{\sqrt{AB^2 - BC^2}}{BC} = f_x - p\cdot\frac{\sqrt{f_x^2 - p'^2}}{p'} \tag{4.14}$$

由 $p' \ll f_x$ 可得

$$FB = f_x - p\cdot\frac{f_x}{p'} = f_x - \frac{f_x}{1+\varepsilon_y} \tag{4.15}$$

由式(4.13)和式(4.15)相等可得

$$\varepsilon_y = \frac{p}{h_y - p} \tag{4.16}$$

由 $p \ll h_y$ 可得

$$\varepsilon_y = \frac{p}{h_y} \tag{4.17}$$

第二步将参考栅和试件栅的栅线平行于 y 方向放置，设其节距均为 p，变形后试件栅的节距为 p'。ε_y 仅使试件栅的栅线沿 y 方向伸长或缩短，θ_x 仅使试件栅的栅线沿 y 方向移动，因此 ε_y 和 θ_x 对云纹条纹没有影响。ε_x 使试件栅的节距增大或减小，θ_y 使试件栅的栅线转动，因此 ε_x 和 θ_y 将产生云纹条纹。同理，可推得

$$\theta_y = \frac{p}{f_y}, \quad \varepsilon_x = \frac{p}{h_x} \tag{4.18}$$

式中，f_y 为云纹条纹在 y 方向(即栅线方向)的间距；h_x 为云纹条纹在 x 方向(即垂直栅线方向)的间距。综合式(4.12)、式(4.17)和式(4.18)，得平面应变

$$\varepsilon_x = \frac{p}{h_x}, \quad \varepsilon_y = \frac{p}{h_y}, \quad \gamma_{xy} = \theta_x + \theta_y = \frac{p}{f_x} + \frac{p}{f_y} \tag{4.19}$$

式中，p 已知，只要测得 h_x、h_y、f_x 和 f_y 即可求得 ε_x、ε_y 和 γ_{xy}。

4.1.3 影像云纹离面位移测量

几何云纹还可用于物体离面位移测量。当几何云纹用于离面位移测量时，称为影像云纹。在影像云纹中，试件栅并不是独立光栅，而是参考栅在光线照射下投射于物体表面而形成的参考栅的影像，其形状随物体表面的高低起伏而变化。

将参考栅放置于物体前面，设照明和观察方向与参考栅法线的夹角分别为 α 和 β，如图 4.7 所示。

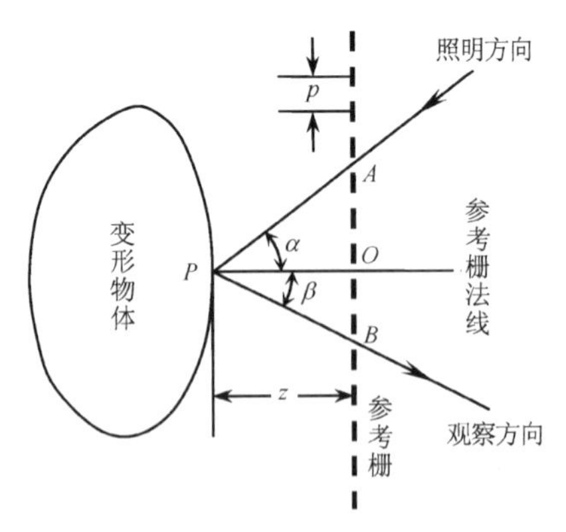

图 4.7 影像云纹

照明光线透过参考栅上的 A 点射到物面上的 P 点，P 点为参考栅 A 点的影像。如果 A

点为透光部分，则 P 点为亮点。沿观察方向看 P 点，看到该点与参考栅上的 B 点重合，若 B 点为透光部分，则得到一系列亮点组成的亮条纹。

由图 4.7 有 $AB = np$，其中 p 为参考栅节距，n 为亮云纹条纹级数 ($n = 0, 1, 2, \cdots$)。设参考栅到物面上 P 点的距离 $OP = z$，则 $AB = z(\tan\alpha + \tan\beta)$。由此得 $np = z(\tan\alpha + \tan\beta)$，即

$$z = \frac{np}{\tan\alpha + \tan\beta} \tag{4.20}$$

由式(4.20)可见，如果 $\alpha \neq 0$，$\beta \neq 0$，当 $n = 0$ 时，有 $z = 0$，这表明物面上与参考栅接触处的点对应 0 级云纹条纹。

采用影像云纹测量物体离面位移通常有四种系统：①平行照射与平行接收；②发散照射与汇聚接收；③平行照射与汇聚接收；④发散照射与平行接收。下面仅对两种常用系统进行分析。

1. 平行照射与平行接收

如图 4.8 所示，平行入射光与参考栅法线的夹角为 α，平行反射光垂直于参考栅，即 $\beta = 0$，因此，式(4.20)可表示为

$$z = \frac{np}{\tan\alpha} \tag{4.21}$$

因此，云纹条纹级数确定后，根据已知的 p 和 α，即可求出物面各点到参考栅的距离。

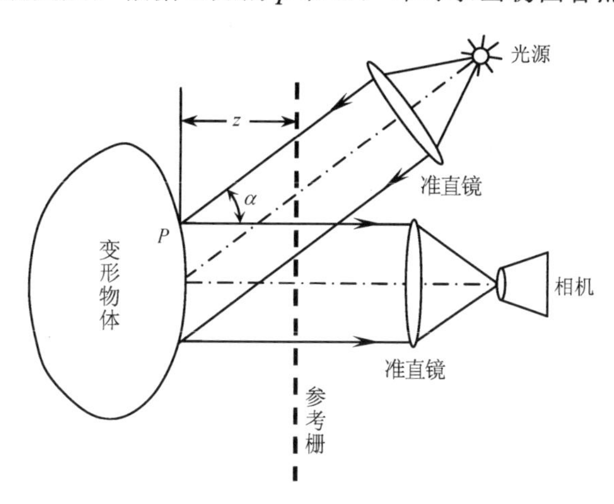

图 4.8　平行照射与平行接收

在物体变形前后分别记录两张云纹条纹图，并进行相减处理，即可得到物体的离面位移场。设物体变形前后分别有 $z_0 = n_0 p / \tan\alpha$ 和 $z = np / \tan\alpha$。记录两张云纹条纹图后做相减处理，得离面位移

$$w = z - z_0 = \frac{np}{\tan\alpha} - \frac{n_0 p}{\tan\alpha} = \frac{(n - n_0)p}{\tan\alpha} \tag{4.22}$$

因此，两张云纹条纹图上对应物面某点的条纹级数确定后，即可通过式(4.22)确定物面上该点的离面位移。

2. 发散照射与汇聚接收

采用平行照射与平行接收，需要光源或相机与物体间的距离比物体尺寸大得多，或者需要在光源与物体间，以及相机与物体间放置孔径大小与物体尺寸相当的透镜，这样就限制了物体的尺寸不能太大。采用发散照射与汇聚接收，光源和相机与物体间可相距有限距离，物体尺寸大小只受覆盖其上的参考栅尺寸的限制。

发散照射与汇聚接收系统如图 4.9 所示，由图得，$\tan\alpha=(L-x)/(D+z)$，$\tan\beta=x/(D+z)$，代入式(4.20)，得 $z=np(D+z)/L$，解出 z，得

$$z=\frac{npD}{L-np} \tag{4.23}$$

考虑到 $np<<L$，则式(4.23)可表示为

$$z=\frac{npD}{L} \tag{4.24}$$

云纹条纹级数确定后，根据已知的 p、D 和 L，即可求出物面各点到参考栅的距离。

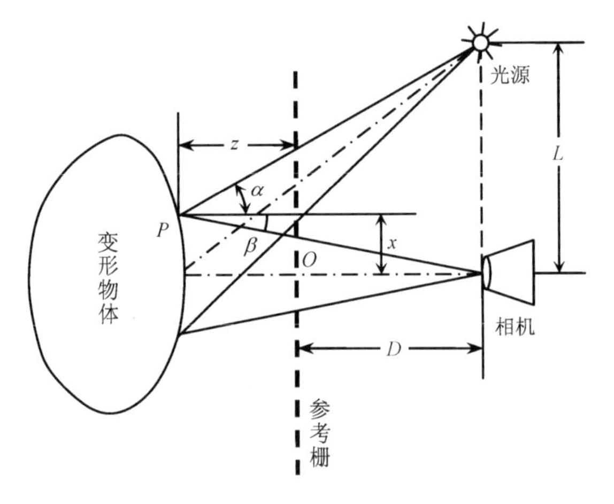

图 4.9　发散照射与汇聚接收系统

在物体变形前后分别记录两张云纹条纹图，并进行相减处理，可得离面位移

$$w=\frac{(n-n_0)pD}{L} \tag{4.25}$$

4.1.4　反射云纹斜率测量

采用反射云纹可以直接测量斜率(即离面位移导数)，将斜率分布曲线再进行一次求导，即可求得弯曲板的曲率和扭率，进而得到弯曲板的弯矩和扭矩。

反射云纹测量系统如图 4.10 所示。光照射反射栅，经反射栅反射后再照射具有反射能力的弯曲板。板变形前，相机记录的是经板上 P 点反射的反射栅上 A 处的栅线；板变形后，记录的是反射栅上 B 处的栅线。假设能够记录到第 n 级亮云纹条纹，则此时有 $AB=np$，其中 p 为反射栅节距，n 为亮云纹条纹级数 $(n=0,1,2,\cdots)$。设反射栅到板的距离为 D，则有 $AB=D[\tan(\alpha+2\theta)-\tan\alpha]$。由此得 $np=D[\tan(\alpha+2\theta)-\tan\alpha]$，经过推导，得

$$\tan 2\theta = \frac{np}{D(1+\tan^2\alpha)+np} \tag{4.26}$$

考虑到 $\tan\alpha = x/D << 1$，且 $np << D$，式(4.26)可简化为

$$\tan 2\theta = \frac{np}{D} \tag{4.27}$$

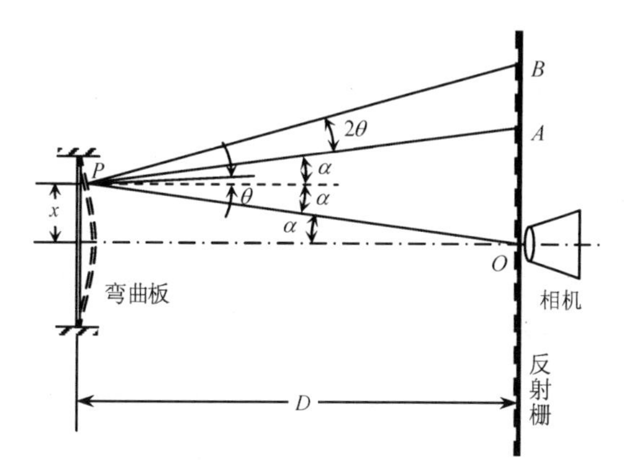

图 4.10　反射云纹测量系统

对于小变形，有 $\tan 2\theta \approx 2\partial w/\partial x$，则

$$\frac{\partial w}{\partial x} = \frac{np}{2D} \tag{4.28}$$

把板旋转 90°，同理可得

$$\frac{\partial w}{\partial y} = \frac{np}{2D} \tag{4.29}$$

式(4.28)和式(4.29)中的 p 和 D 为已知，因此只要测得云纹条纹级数 n，即可由式(4.28)和式(4.29)求出斜率 $\partial w/\partial x$ 和 $\partial w/\partial y$。

4.2　云 纹 干 涉

云纹干涉由于采用高密度衍射光栅作为试件栅，其测量灵敏度与全息干涉或散斑干涉相同。

4.2.1　实时法面内位移测量

面内位移测量系统如图 4.11 所示。当两束对称入射平行光波的入射角满足 $\alpha = \arcsin(\lambda/p)$ 时，则在试件表面法线方向分别获得衍射光波 O_1 和 O_2。若试件栅十分规整，试件也未受力，则两个衍射光波 O_1 和 O_2 为平面波，可分别表示为

$$O_1 = a\exp(\mathrm{i}\varphi_1), \quad O_2 = a\exp(\mathrm{i}\varphi_2) \tag{4.30}$$

式中，a 为光波振幅；φ_1 和 φ_2 分别为两光波相位(对于平面波，φ_1 和 φ_2 均为常数)。

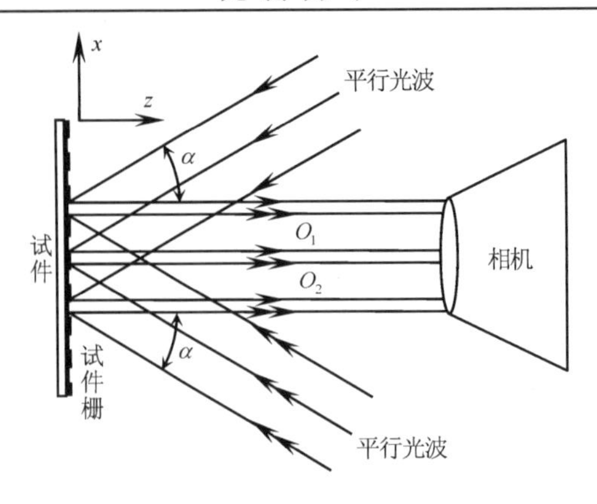

图 4.11　面内位移测量系统

试件受力发生变形，则平面波 O_1 和 O_2 变为与试件表面变形有关的翘曲波前，其相位将发生相应的变化。翘曲波前 O_1' 和 O_2' 可分别表示为

$$O_1' = a\exp[\mathrm{i}(\varphi_1 + \delta_1)], \quad O_2' = a\exp[\mathrm{i}(\varphi_2 + \delta_2)] \tag{4.31}$$

式中，δ_1 和 δ_2 为由于试件表面变形而引起的相位变化。相位变化 δ_1 和 δ_2 与 x 和 z 方向的位移 $u(x,y)$ 和 $w(x,y)$ 有如下关系：

$$\delta_1 = k[w(1+\cos\alpha) + u\sin\alpha], \quad \delta_2 = k[w(1+\cos\alpha) - u\sin\alpha] \tag{4.32}$$

式中，$w = w(x,y)$ 和 $u = u(x,y)$ 分别是离面位移和沿 x 方向的面内位移。两束翘曲光波经过成像系统后在像面发生干涉，像面所记录的强度分布可表示为

$$I = (O_1' + O_2')(O_1' + O_2')^* = 2a^2[1+\cos(\varphi+\delta)] \tag{4.33}$$

式中，$\varphi = \varphi_1 - \varphi_2$ 为两束平面波 O_1 和 O_2 的初始相位差，为一常数，可等效为试件平移所产生的均匀相位；$\delta = \delta_1 - \delta_2$ 为试件变形后两束翘曲衍射光波的相对相位变化，根据式(4.32)，可得

$$\delta = 2ku\sin\alpha \tag{4.34}$$

为了获得另一面内位移分量 $v(x,y)$，应使试件栅和对称入射光波旋转90°，则

$$\delta = 2kv\sin\alpha \tag{4.35}$$

为了同时获得两个面内位移分量，通常在试件表面复制正交光栅。正交相位光栅不仅能产生沿 x 和 y 方向的衍射光波，而且还可以产生沿+45°和–45°方向的衍射光波。因此，采用正交光栅可以同时获得沿 x、y、+45°和–45°方向的面内位移分量。

4.2.2　差载法面内位移测量

通常难以获得绝对准确的平面衍射光波，因此上述相位差 $\varphi = \varphi_1 - \varphi_2$ 不是常数。当 φ 不是常数时，需要采用差载方法测量面内位移。为了消除 φ 对位移条纹的影响，可在光波 O_1（或 O_2）光路中附加光楔 $f(x,y)$，则加载前衍射光波可分别表示为

$$O_1 = a\exp[\mathrm{i}(\varphi_1 + f)], \quad O_2 = a\exp(\mathrm{i}\varphi_2) \tag{4.36}$$

相应的强度分布为

$$I_1=(O_1+O_2)(O_1+O_2)^*=2a^2[1+\cos(\varphi+f)] \tag{4.37}$$

式中，$\varphi=\varphi_1-\varphi_2$。

加载后衍射光波可表示为

$$O_1'=a\exp[\mathrm{i}(\varphi_1+\delta_1+f)],\quad O_2'=a\exp[\mathrm{i}(\varphi_2+\delta_2)] \tag{4.38}$$

式中，$\delta_1=k[w(1+\cos\alpha)+u\sin\alpha]$；$\delta_2=k[w(1+\cos\alpha)-u\sin\alpha]$。相应的强度分布为

$$I_2=(O_1'+O_2')(O_1'+O_2')^*=2a^2[1+\cos(\varphi+\delta+f)] \tag{4.39}$$

式中，$\delta=\delta_1-\delta_2=2ku\sin\alpha$。

两次曝光的总强度分布为

$$I=I_1+I_2=4a^2\left[1+\cos\left(\varphi+\frac{1}{2}\delta+f\right)\cos\left(\frac{1}{2}\delta\right)\right] \tag{4.40}$$

两次曝光照相底片经显影和定影处理后，置于滤波系统中，即可获得符合下列条件的暗条纹：

$$\cos\left(\frac{1}{2}\delta\right)=0 \tag{4.41}$$

即

$$\delta=(2n+1)\pi\quad(n=0,\pm1,\pm2,\cdots) \tag{4.42}$$

把式(4.34)代入式(4.42)，得

$$u=\frac{(2n+1)\pi}{2k\sin\alpha}=\frac{(2n+1)\lambda}{4\sin\alpha}\quad(n=0,\pm1,\pm2,\cdots) \tag{4.43}$$

第5章　相移干涉与相位展开

在光测力学中，待测量(如位移、应变等)与干涉条纹所包含的相位信息几乎直接相关，因此从干涉条纹提取相位信息就显得极其重要。全息干涉、散斑干涉和云纹干涉等光测技术都是记录相干光波因相互干涉而形成的干涉条纹。干涉条纹表示相位等值线，即同一条纹中心线上各点具有相同的相位值，相邻条纹中心线之间具有相同的相位差。

传统相位检测方法需要进行条纹定位和级数确定，以便得到干涉条纹图的相位分布。但传统方法往往会引起较大的测量误差，原因有二：一是亮度极值位置未必就处在条纹中心线上；二是通过插值才能确定相邻条纹之间各点的相位。为了弥补传统相位检测方法的不足，人们提出了多种相位检测方法。

在光测力学中，常用的相位检测方法是相移干涉。相移干涉不需要进行条纹定位和级数确定，即可直接得到干涉条纹图的相位分布。

5.1　相 移 干 涉

根据引入相移方式的不同，相移干涉分为时间相移干涉和空间相移干涉。时间相移干涉是指在时间序列上采集图像，在各帧图像之间形成固定的相位差。空间相移干涉是指在空间序列上采集图像，在空间不同位置之间形成固定的相位差。

5.1.1　时间相移干涉

1. 时间相移原理

在光测力学中，两束相干光波相互干涉而在记录面上形成的强度分布可表示为

$$I(x,y)=I_0(x,y)[1+V(x,y)\cos\delta(x,y)] \tag{5.1}$$

式中，$I_0(x,y)$为背景强度；$V(x,y)$为条纹可见度(或调制度)；$\delta(x,y)$为待测相位。

当干涉条纹图通过CCD(charge coupled device)或CMOS(complementary metal oxide semiconductor)采样并量化为数字图像时，电子噪声和散斑噪声等都会影响干涉条纹图的强度分布。因此综合考虑这些因素之后，干涉条纹图的强度分布可表示为

$$I(x,y)=A(x,y)+B(x,y)\cos\delta(x,y) \tag{5.2}$$

式中，$A(x,y)$和$B(x,y)$分别为干涉条纹图的背景强度和调制强度。式(5.2)中的$I(x,y)$是已知量，但$A(x,y)$、$B(x,y)$和$\delta(x,y)$均为未知量，即上述一个方程含有3个未知量，因此若上述方程可解，则至少要有3个独立方程，才能确定待测相位$\delta(x,y)$。

时间相移干涉在时间序列上采集图像，在各幅图像之间形成已知相位差，通过采集至

少 3 幅图像，即可联解方程组而得到待测相位分布。设第 n 幅干涉条纹图的相移量为 α_n，则干涉条纹图的强度分布为

$$I_n(x,y)=A(x,y)+B(x,y)\cos[\delta(x,y)+\alpha_n] \quad (n=1,\ 2,\ \cdots,\ N;\ N\geqslant 3) \tag{5.3}$$

式中，$I_n(x,y)$ 和 α_n 为已知量，只有 $A(x,y)$、$B(x,y)$ 和 $\delta(x,y)$ 为未知量，因此通过引进不同的相移量 α_n，构造至少 3 个方程，从而确定待测相位 $\delta(x,y)$。

2. 时间相移算法

时间相移干涉主要包括三步算法、四步算法和 Carré 算法等。这些算法已经广泛应用于光测力学。

1) 三步算法

设三次相移量依次为 α_1、α_2、α_3，则 3 幅干涉条纹图的强度分布可表示为

$$\begin{aligned}I_1(x,y)&=A(x,y)+B(x,y)\cos[\delta(x,y)+\alpha_1]\\I_2(x,y)&=A(x,y)+B(x,y)\cos[\delta(x,y)+\alpha_2]\\I_3(x,y)&=A(x,y)+B(x,y)\cos[\delta(x,y)+\alpha_3]\end{aligned} \tag{5.4}$$

联立求解，得到干涉条纹图的相位分布为

$$\frac{\tan\delta(x,y)(\sin\alpha_2-\sin\alpha_3)-(\cos\alpha_2-\cos\alpha_3)}{\tan\delta(x,y)(2\sin\alpha_1-\sin\alpha_2-\sin\alpha_3)-(2\cos\alpha_1-\cos\alpha_2-\cos\alpha_3)}=\frac{I_2(x,y)-I_3(x,y)}{2I_1(x,y)-I_2(x,y)-I_3(x,y)} \tag{5.5}$$

式(5.5)是三步算法的一般表达式。其特例如下。

(1) 如果三次相移量依次为 0、π/3 和 2π/3（相移增量为 π/3），则干涉条纹图的相位分布为

$$\delta(x,y)=\arctan\frac{2I_1(x,y)-3I_2(x,y)+I_3(x,y)}{\sqrt{3}[I_2(x,y)-I_3(x,y)]} \tag{5.6}$$

(2) 如果三次相移量依次为 0、π/2 和 π（相移增量为 π/2），则干涉条纹图的相位分布为

$$\delta(x,y)=\arctan\frac{I_1(x,y)-2I_2(x,y)+I_3(x,y)}{I_1(x,y)-I_3(x,y)} \tag{5.7}$$

(3) 如果三次相移量依次为 0、2π/3 和 4π/3（相移增量为 2π/3），则干涉条纹图的相位分布为

$$\delta(x,y)=\arctan\frac{\sqrt{3}[I_3(x,y)-I_2(x,y)]}{2I_1(x,y)-I_2(x,y)-I_3(x,y)} \tag{5.8}$$

2) 四步算法

设四次相移量依次为 α_1、α_2、α_3、α_4，则 4 幅干涉条纹图的强度分布可表示为

$$\begin{aligned}I_1(x,y)&=A(x,y)+B(x,y)\cos[\delta(x,y)+\alpha_1]\\I_2(x,y)&=A(x,y)+B(x,y)\cos[\delta(x,y)+\alpha_2]\\I_3(x,y)&=A(x,y)+B(x,y)\cos[\delta(x,y)+\alpha_3]\\I_4(x,y)&=A(x,y)+B(x,y)\cos[\delta(x,y)+\alpha_4]\end{aligned} \tag{5.9}$$

联立求解，得干涉条纹图的相位分布为

$$\frac{\tan\delta(x,y)(\sin\alpha_1-\sin\alpha_3)-(\cos\alpha_1-\cos\alpha_3)}{\tan\delta(x,y)(\sin\alpha_2-\sin\alpha_4)-(\cos\alpha_2-\cos\alpha_4)}=\frac{I_1(x,y)-I_3(x,y)}{I_2(x,y)-I_4(x,y)} \tag{5.10}$$

式(5.10)是四步算法的一般表达式。其特例如下。

(1)如果四次相移量依次为π/4、3π/4、5π/4和7π/4(相移增量为π/2)，则干涉条纹图的相位分布为

$$\delta(x,y)=\arctan\frac{[I_2(x,y)-I_4(x,y)]+[I_1(x,y)-I_3(x,y)]}{[I_2(x,y)-I_4(x,y)]-[I_1(x,y)-I_3(x,y)]} \tag{5.11}$$

(2)如果四次相移量依次为0、π/3、2π/3和π(相移增量为π/3)，则干涉条纹图的相位分布为

$$\delta(x,y)=\arctan\frac{I_1(x,y)-I_2(x,y)-I_3(x,y)+I_4(x,y)}{\sqrt{3}[I_2(x,y)-I_3(x,y)]} \tag{5.12}$$

(3)如果四次相移量依次为0、π/2、π和3π/2(相移增量为π/2)，则干涉条纹图的相位分布为

$$\delta(x,y)=\arctan\frac{I_4(x,y)-I_2(x,y)}{I_1(x,y)-I_3(x,y)} \tag{5.13}$$

3) Carré 算法

如果四次相移量依次为-3α、$-\alpha$、α和3α(相移增量为2α，但α未知)，则 4 幅干涉条纹图的强度分布可表示为

$$\begin{aligned}I_1(x,y)&=A(x,y)+B(x,y)\cos[\delta(x,y)-3\alpha]\\I_2(x,y)&=A(x,y)+B(x,y)\cos[\delta(x,y)-\alpha]\\I_3(x,y)&=A(x,y)+B(x,y)\cos[\delta(x,y)+\alpha]\\I_4(x,y)&=A(x,y)+B(x,y)\cos[\delta(x,y)+3\alpha]\end{aligned} \tag{5.14}$$

联立求解，得到干涉条纹图的相位分布为

$$\delta(x,y)=\arctan\left\{\tan\beta\frac{[I_2(x,y)-I_3(x,y)]+[I_1(x,y)-I_4(x,y)]}{[I_2(x,y)+I_3(x,y)]-[I_1(x,y)+I_4(x,y)]}\right\} \tag{5.15}$$

式中，β可通过下式得到

$$\tan^2\beta=\frac{3[I_2(x,y)-I_3(x,y)]-[I_1(x,y)-I_4(x,y)]}{[I_2(x,y)-I_3(x,y)]+[I_1(x,y)-I_4(x,y)]} \tag{5.16}$$

3. 时间相移实验

图 5.1 所示为采用三步相移干涉而得到的干涉条纹图。其中，图 5.1(a)、图 5.1(b)和图 5.1(c)的相移量依次为0、π/2和π。

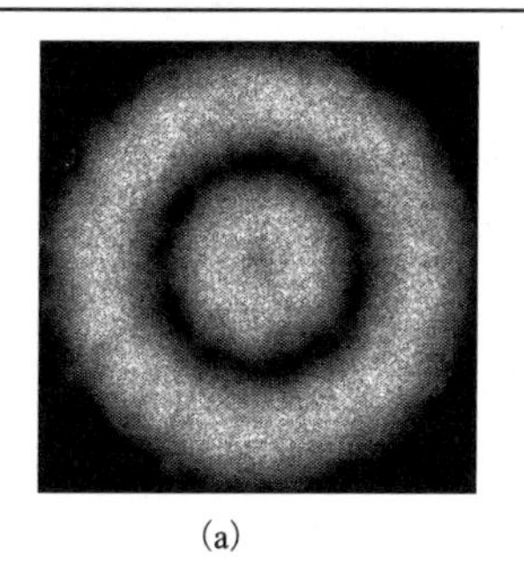
(a)

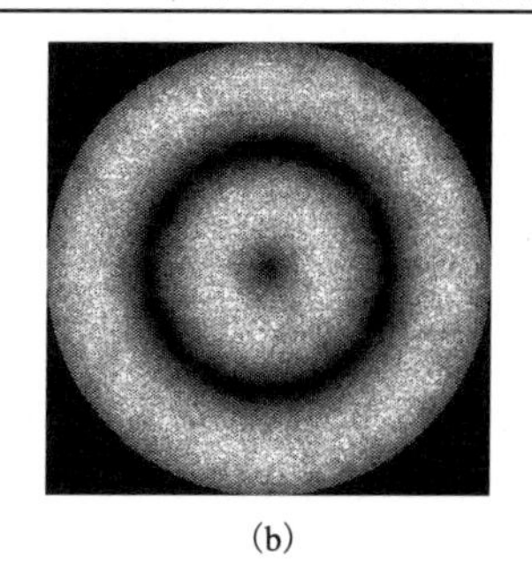
(b)

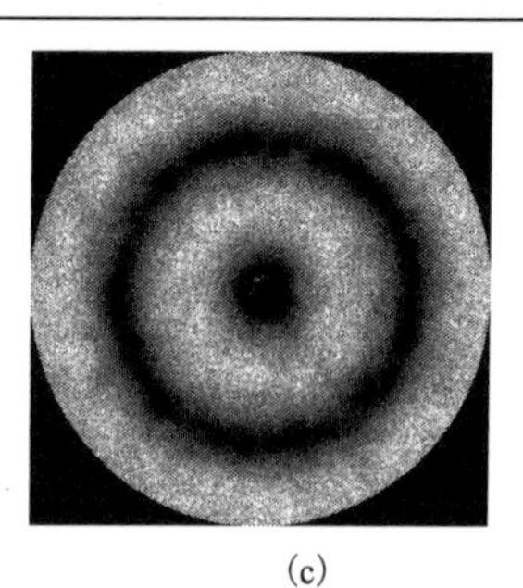
(c)

图 5.1　干涉条纹图

图 5.2 所示为上述 3 幅干涉条纹图通过三步算法而得到的包裹相位分布图，相位分布区间为 $-\pi/2 \sim \pi/2$。

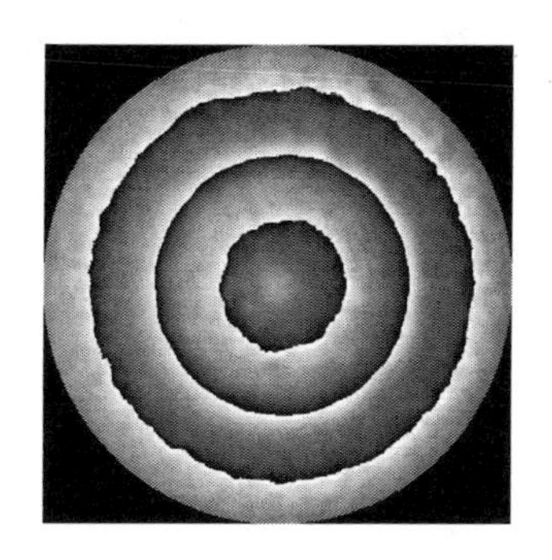

图 5.2　包裹相位分布图

5.1.2　空间相移干涉

1．空间相移原理

在空间相移干涉中，通常采用的方法是引入载波，因此空间相移干涉也称为空间载波法。空间载波法通过一幅干涉条纹图就可以得到相位分布。

引入空间载波后，干涉条纹图所记录的强度分布可表示为

$$I(x,y)=A(x,y)+B(x,y)\cos[\delta(x,y)+2\pi fx] \tag{5.17}$$

式中，$f=f(x)$ 为沿 x 方向(载波方向)所施加的线性空间载波；$\delta(x,y)$ 为待测相位。

干涉条纹图由 CCD 记录并存储为数字图像，其第 (i,j) 像素记录的强度为

$$I(x_i,y_j)=A(x_i,y_j)+B(x_i,y_j)\cos[\delta(x_i,y_j)+2\pi fx_i] \tag{5.18}$$

式中，$i=1,2,\cdots,M;\ j=1,2,\cdots,N$，其中 $M\times N$ 为 CCD 的像素。

2．空间相移算法

1) 三步算法

假设第 (i,j)、$(i+1,j)$ 和 $(i+2,j)$ 等相邻像素具有相同的背景强度、条纹对比度和待测相位，则第 (i,j)、$(i+1,j)$ 和 $(i+2,j)$ 像素的强度分别为

$$
\begin{aligned}
I(x_i, y_j) &= A(x_i, y_j) + B(x_i, y_j)\cos[\delta(x_i, y_j) + 2\pi f x_i] \\
I(x_{i+1}, y_j) &= A(x_i, y_j) + B(x_i, y_j)\cos[\delta(x_i, y_j) + 2\pi f (x_i + \Delta x)] \\
I(x_{i+2}, y_j) &= A(x_i, y_j) + B(x_i, y_j)\cos[\delta(x_i, y_j) + 2\pi f (x_i + 2\Delta x)]
\end{aligned} \tag{5.19}
$$

式中，Δx 为 CCD 沿 x 方向(载波方向)相邻像素的中心间距；$2\pi f x_i$ 为第 (i, j) 像素处的载波相位，其中，$i = 1,2,\cdots, M-2; j = 1,2,\cdots,N$。

采用三步算法时，载波频率选择如下。相邻条纹中心间距等于 CCD 相邻像素中心间距的 3 倍。因此，沿 x 方向(载波方向)的相邻像素之间由载波引入的相位差为 $2\pi/3$，即 $2\pi f\Delta x = 2\pi/3$，则式(5.19)可重写为

$$
\begin{aligned}
I(x_i, y_j) &= A(x_i, y_j) + B(x_i, y_j)\cos[\delta(x_i, y_j) + 2\pi f x_i] \\
I(x_{i+1}, y_j) &= A(x_i, y_j) + B(x_i, y_j)\cos\left[\delta(x_i, y_j) + 2\pi f x_i + \frac{2}{3}\pi\right] \\
I(x_{i+2}, y_j) &= A(x_i, y_j) + B(x_i, y_j)\cos\left[\delta(x_i, y_j) + 2\pi f x_i + \frac{4}{3}\pi\right]
\end{aligned} \tag{5.20}
$$

联立求解，得到相位表达式为

$$
\delta(x_i, y_j) = \arctan\left\{\frac{\sqrt{3}[I(x_{i+2}, y_j) - I(x_{i+1}, y_j)]}{2I(x_i, y_j) - I(x_{i+1}, y_j) - I(x_{i+2}, y_j)}\right\} - 2\pi f x_i \tag{5.21}
$$

通过式(5.21)即可求得干涉条纹图各点的相位分布。

2) 四步算法

假设第 (i, j)、$(i+1, j)$、$(i+2, j)$ 和 $(i+3, j)$ 等相邻像素具有相同的背景强度、条纹对比度和待测相位，则第 (i, j)、$(i+1, j)$、$(i+2, j)$ 和 $(i+3, j)$ 像素的强度分别为

$$
\begin{aligned}
I(x_i, y_j) &= A(x_i, y_j) + B(x_i, y_j)\cos[\delta(x_i, y_j) + 2\pi f x_i] \\
I(x_{i+1}, y_j) &= A(x_i, y_j) + B(x_i, y_j)\cos[\delta(x_i, y_j) + 2\pi f (x_i + \Delta x)] \\
I(x_{i+2}, y_j) &= A(x_i, y_j) + B(x_i, y_j)\cos[\delta(x_i, y_j) + 2\pi f (x_i + 2\Delta x)] \\
I(x_{i+3}, y_j) &= A(x_i, y_j) + B(x_i, y_j)\cos[\delta(x_i, y_j) + 2\pi f (x_i + 3\Delta x)]
\end{aligned} \tag{5.22}
$$

式中，$i = 1,2,\cdots,M-3; j = 1,2,\cdots,N$。

采用四步算法时，载波频率按下述要求进行选择：相邻条纹中心间距等于 CCD 相邻像素中心间距的 4 倍，即沿 x 方向(载波方向)的相邻像素之间由载波引入的相位差为 $\pi/2$ 或 $2\pi f\Delta x = \pi/2$，则由式(5.22)得

$$
\begin{aligned}
I(x_i, y_j) &= A(x_i, y_j) + B(x_i, y_j)\cos[\delta(x_i, y_j) + 2\pi f x_i] \\
I(x_{i+1}, y_j) &= A(x_i, y_j) + B(x_i, y_j)\cos\left[\delta(x_i, y_j) + 2\pi f x_i + \frac{1}{2}\pi\right] \\
I(x_{i+2}, y_j) &= A(x_i, y_j) + B(x_i, y_j)\cos[\delta(x_i, y_j) + 2\pi f x_i + \pi] \\
I(x_{i+3}, y_j) &= A(x_i, y_j) + B(x_i, y_j)\cos\left[\delta(x_i, y_j) + 2\pi f x_i + \frac{3}{2}\pi\right]
\end{aligned} \tag{5.23}
$$

联立求解，得

$$
\delta(x_i, y_j) = \arctan\left[\frac{I(x_{i+3}, y_j) - I(x_{i+1}, y_j)}{I(x_i, y_j) - I(x_{i+2}, y_j)}\right] - 2\pi f x_i \tag{5.24}
$$

通过式(5.24)同样可确定干涉条纹图各点的相位分布。

显然，无论是三步算法还是四步算法，空间载波法只需一幅载波干涉条纹图即可得到全场相位分布信息。

3. 空间相移实验

图 5.3 所示为空间载波实验结果，其中图 5.3(a)是调制干涉条纹分布，图 5.3(b)是调制包裹相位分布。

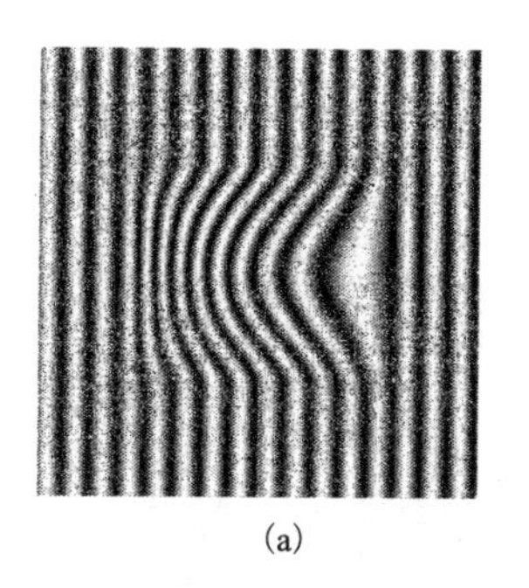

(a)

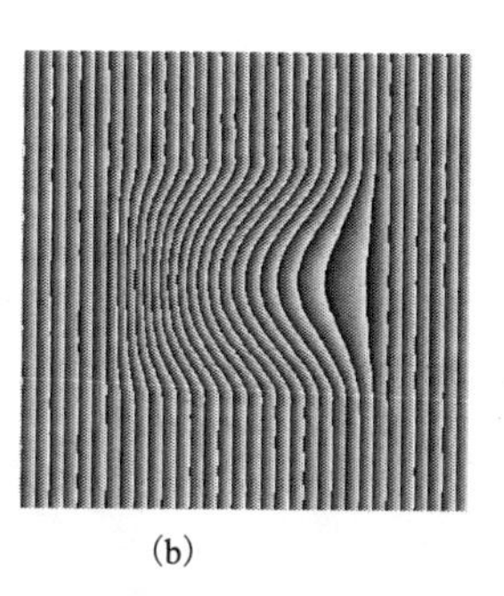

(b)

图 5.3　空间载波实验结果

5.2 相 位 展 开

1. 相位展开原理

采用相移干涉，所得到的相位分布可表示为

$$\delta(x,y)=\arctan\frac{S(x,y)}{C(x,y)} \tag{5.25}$$

式(5.25)表示的相位分布处于 $-\pi/2\sim\pi/2$ 的范围，即 $\delta(x,y)$ 是位于 $-\pi/2\sim\pi/2$ 范围内的包裹相位。

根据 $S(x,y)$ 和 $C(x,y)$ 的正负号，式(5.25)所表示的位于 $-\pi/2\sim\pi/2$ 范围内的包裹相位可以通过如下变换扩展到 $0\sim2\pi$ 的范围：

$$\delta(x,y)=\begin{cases}\delta(x,y) & (S(x,y)\geqslant0,\ C(x,y)>0)\\ \dfrac{1}{2}\pi & (S(x,y)>0,\ C(x,y)=0)\\ \delta(x,y)+\pi & (C(x,y)<0)\\ \dfrac{3}{2}\pi & (S(x,y)<0,\ C(x,y)=0)\\ \delta(x,y)+2\pi & (S(x,y)<0,\ C(x,y)>0)\end{cases} \tag{5.26}$$

经过相位扩展，相位分布区间已由 $-\pi/2 \sim \pi/2$ 变为 $0 \sim 2\pi$，此时所得到的相位分布是位于 $0 \sim 2\pi$ 范围内的包裹相位。

显然，利用式(5.26)经过相位扩展后所得到的相位分布 $\delta(x,y)$ 仍然是包裹相位，因此要得到连续相位分布则需要对包裹相位 $\delta(x,y)$ 进行相位展开。利用二维相位展开算法，如果相邻像素之间的相位差达到或超过 π，则通过增加或减少 2π 的整数倍相位，就可消除相位的不连续性。展开相位与位于 $0 \sim 2\pi$ 范围的包裹相位之间的关系可表示为

$$\delta_{\mathrm{u}}(x,y) = \delta(x,y) + 2\pi n(x,y) \tag{5.27}$$

式中，$n(x,y)$ 为整数。

2. 相位展开实验

实验一　图 5.4 所示为时间相移干涉相位分布图，其中图 5.4(a)和图 5.4(b)为包裹相位图，其相位区间分别为 $-\pi/2 \sim \pi/2$ 和 $0 \sim 2\pi$，图 5.4(c)为连续相位图或展开相位图。

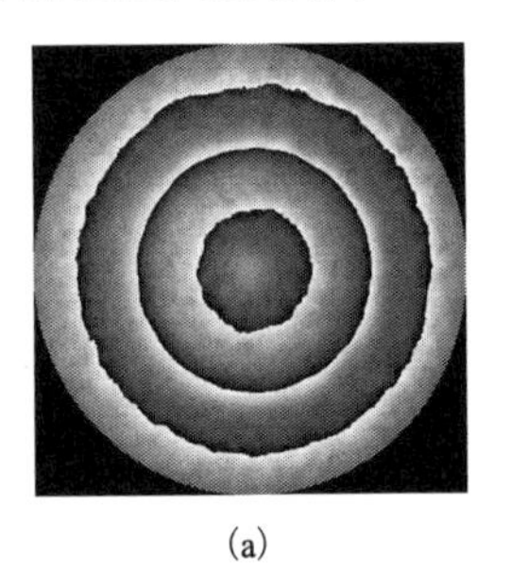
(a)

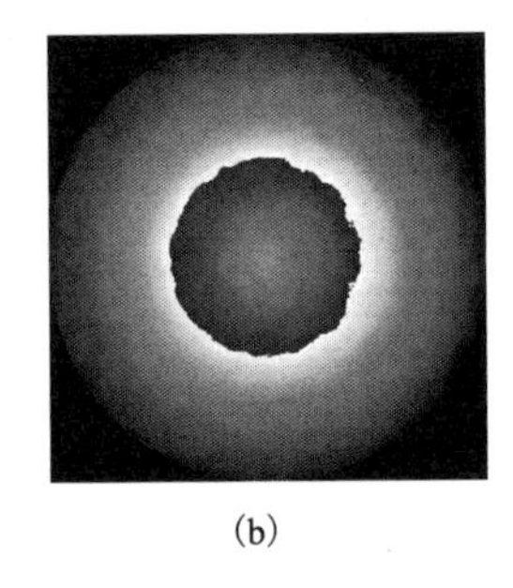
(b)

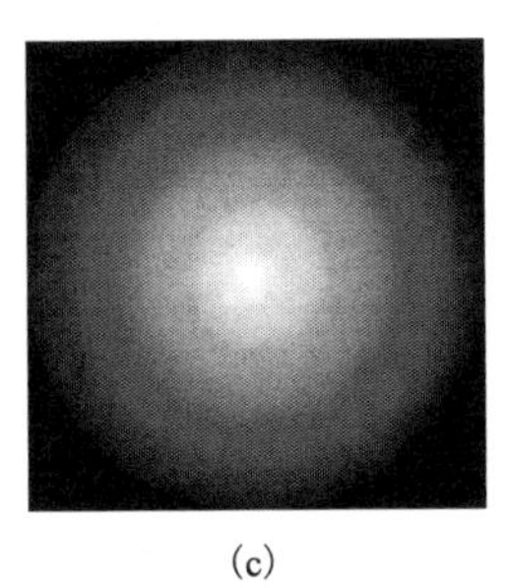
(c)

图 5.4　时间相移干涉相位分布图

实验二　图 5.5 所示为空间载波相位分布图，其中图 5.5(a)是调制包裹相位分布，图 5.5(b)是调制连续相位分布，图 5.5(c)是变形连续相位分布。

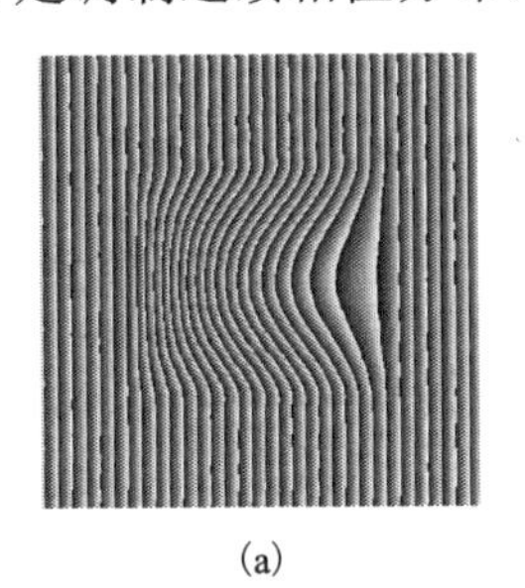
(a)

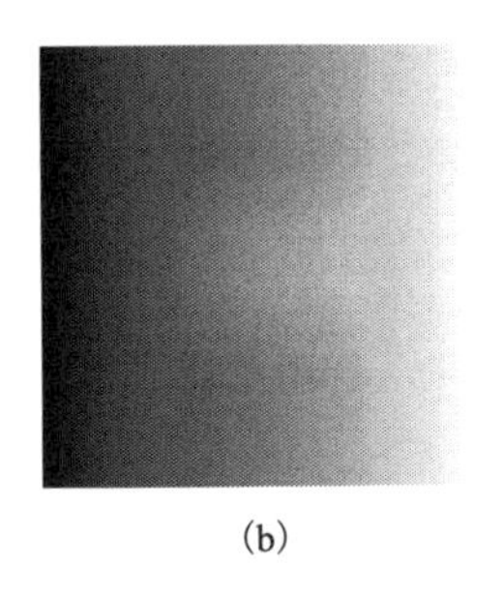
(b)

(c)

图 5.5　空间载波相位分布图

第 6 章　离散变换与低通滤波

在光测力学中，尤其是在数字散斑干涉中，所得到的条纹往往含有严重的噪声。为了得到表征物体变形的连续相位分布，需要选用合适的滤波方法对条纹进行降噪。

条纹降噪可以在空域进行，也可以在频域进行。如果在频域进行，则条纹图需要从空域变换到频域。

6.1　离 散 变 换

用于光测力学的常用变换方法主要包括离散傅里叶变换和离散余弦变换等。当图像阵列很大时，若直接在空域进行处理，计算量将会很大。为了减小计算量，可以采用频域处理方法，即通过图像变换，如离散傅里叶变换和离散余弦变换等，将图像从空域变换到频域，在频域对图像进行处理，处理结果再从频域反变换到空域，进而得到所需要的图像。

6.1.1　离散傅里叶变换

离散傅里叶变换描述了离散信号的空域表示与频域表示之间的关系，是信号处理的有效工具之一，对频谱分析、卷积与相关运算、滤波处理、功率谱分析和传递函数建模等的快速计算起到关键作用。利用离散傅里叶变换的空域与频域分析方法可解决很多图像处理问题，因而离散傅里叶变换在数字图像处理领域具有广泛的应用。

1. 离散傅里叶变换原理

1) 一维离散傅里叶变换

设 $f(m)$ 是在时域上等间隔采样得到的 M 点离散信号，其中，$m=0,1,\cdots,M-1$，则一维离散傅里叶变换定义为

$$F(p)=\sum_{m=0}^{M-1} f(m)\exp\left(-\mathrm{i}2\pi\frac{mp}{M}\right) \quad (p=0,1,\cdots,M-1) \tag{6.1}$$

式中，p 为频域像素坐标；$\exp\left(-\mathrm{i}2\pi\frac{mp}{M}\right)$ 为变换核。

一维离散傅里叶反变换定义为

$$f(m)=\frac{1}{M}\sum_{p=0}^{M-1} F(p)\exp\left(\mathrm{i}2\pi\frac{mp}{M}\right) \quad (m=0,1,\cdots,M-1) \tag{6.2}$$

式中，$\exp\left(\mathrm{i}2\pi\frac{mp}{M}\right)$ 为反变换核。

2) 二维离散傅里叶变换

设 $f(m,n)$ 是在空域上等间隔采样得到的 $M\times N$ 二维离散信号，则二维离散傅里叶变换定义为

$$F(p,q)=\sum_{m=0}^{M-1}\sum_{n=0}^{N-1}f(m,n)\exp\left[-\mathrm{i}2\pi\left(\frac{mp}{M}+\frac{nq}{N}\right)\right]\quad(p=0,1,\cdots,M-1;q=0,1,\cdots,N-1)\tag{6.3}$$

式中，(p,q) 为频域像素坐标。

二维离散傅里叶反变换定义为

$$f(m,n)=\frac{1}{MN}\sum_{p=0}^{M-1}\sum_{q=0}^{N-1}F(p,q)\exp\left[\mathrm{i}2\pi\left(\frac{mp}{M}+\frac{nq}{N}\right)\right]\quad(m=0,1,\cdots,M-1;n=0,1,\cdots,N-1)\tag{6.4}$$

2. 快速傅里叶变换原理

在数字图像处理中，当图像阵列较大时，直接采用离散傅里叶变换往往具有很大的计算量。为了减小计算量，人们提出了快速傅里叶变换。快速傅里叶变换就是将离散傅里叶变换的乘法运算转变为加(减)法运算。快速傅里叶变换的提出为离散傅里叶变换的广泛应用奠定了基础。

1) 一维快速傅里叶变换

设变换核表示为

$$W_M^{mp}=\exp\left(-\mathrm{i}2\pi\frac{mp}{M}\right)\tag{6.5}$$

则一维离散傅里叶变换可表示为

$$F(p)=\sum_{m=0}^{M-1}f(m)W_M^{mp}\quad(p=0,1,\cdots,M-1)\tag{6.6}$$

式(6.6)表明，每计算一个 $F(p)$，需要进行 M 次乘法和 $M-1$ 次加法。对 M 个采样点，要进行 M^2 次乘法和 $M(M-1)$ 次加法。当 M 很大时，计算量将非常大。采用快速傅里叶变换算法时，式(6.6)的计算将分两步进行。

第 1 步：考虑在 $(0,1,\cdots,M/2-1)$ 范围的 $F(p)$，则式(6.6)可表示为

$$F(p)=\sum_{m=0}^{M/2-1}f(2m)W_M^{2mp}+\sum_{m=0}^{M/2-1}f(2m+1)W_M^{(2m+1)p}\quad(p=0,1,\cdots,M/2-1)\tag{6.7}$$

利用 $W_M^{2mp}=W_{M/2}^{mp}$，得

$$F(p)=\sum_{m=0}^{M/2-1}f(2m)W_{M/2}^{mp}+W_M^p\sum_{m=0}^{M/2-1}f(2m+1)W_{M/2}^{mp}\quad(p=0,1,\cdots,M/2-1)\tag{6.8}$$

设 $F_{\mathrm{e}}(p)=\sum_{m=0}^{M/2-1}f(2m)W_{M/2}^{mp}$ 和 $F_{\mathrm{o}}(p)=\sum_{m=0}^{M/2-1}f(2m+1)W_{M/2}^{mp}$，则式(6.8)可表示为

$$F(p)=F_{\mathrm{e}}(p)+W_M^pF_{\mathrm{o}}(p)\quad(p=0,1,\cdots,M/2-1)\tag{6.9}$$

式中，$F_e(p)$ 和 $F_o(p)$ 分别为偶数项和奇数项。

第 2 步：考虑在 $(M/2, M/2+1, \cdots, M-1)$ 范围的 $F(p)$，则式(6.6)可表示为

$$F(p+M/2)=\sum_{m=0}^{M/2-1} f(2m)W_M^{2m(p+M/2)}+\sum_{m=0}^{M/2-1} f(2m+1)W_M^{(2m+1)(p+M/2)} \quad (p=0,1,\cdots,M/2-1) \tag{6.10}$$

利用 $W_M^{2mp}=W_{M/2}^{mp}$、$W_M^{mM}=1$ 和 $W_M^{M/2}=-1$，得

$$F(p+M/2)=\sum_{m=0}^{M/2-1} f(2m)W_{M/2}^{mp}-W_M^p\sum_{m=0}^{M/2-1} f(2m+1)W_{M/2}^{mp}=F_e(p)-W_M^pF_o(p) \quad (p=0,1,\cdots,M/2-1) \tag{6.11}$$

上述内容表明，快速傅里叶变换首先将原函数分为偶数项和奇数项，然后不断将偶数项和奇数项相加(减)，进而得到所需要的结果。快速傅里叶变换的步骤是：第 1 步，将 1 个 M 点的离散傅里叶变换转化为 2 个 $M/2$ 点的离散傅里叶变换；第 2 步，将 2 个 $M/2$ 点的离散傅里叶变换转化为 4 个 $M/4$ 点的离散傅里叶变换；依次类推。

2)二维快速傅里叶变换

二维离散傅里叶变换可表示为

$$F(p,q)=\sum_{m=0}^{M-1}\sum_{n=0}^{N-1} f(m,n)W_M^{mp}W_N^{nq} \quad (p=0,1,\cdots,M-1;q=0,1,\cdots,N-1) \tag{6.12}$$

式中，$W_M^{mp}=\exp\left(-\mathrm{i}2\pi\dfrac{mp}{M}\right)$；$W_N^{nq}=\exp\left(-\mathrm{i}2\pi\dfrac{nq}{N}\right)$。式(6.12)还可分别表示为

$$F(p,q)=\sum_{m=0}^{M-1}\left[\sum_{n=0}^{N-1} f(m,n)W_N^{nq}\right]W_M^{mp} \quad (p=0,1,\cdots,M-1;q=0,1,\cdots,N-1) \tag{6.13}$$

或

$$F(p,q)=\sum_{n=0}^{N-1}\left[\sum_{m=0}^{M-1} f(m,n)W_M^{mp}\right]W_N^{nq} \quad (p=0,1,\cdots,M-1;q=0,1,\cdots,N-1) \tag{6.14}$$

上述内容表明，二维快速傅里叶变换可以转化为两次一维快速傅里叶变换，即可先对图像矩阵的各列(或行)取快速傅里叶变换，然后对各行(或列)取快速傅里叶变换。因此经过两次一维快速傅里叶变换，即可实现二维快速傅里叶变换。

3. 离散傅里叶变换实验

离散傅里叶变换应用于数字图像处理的基本思路是将数字图像从空域变换到频域，然后在频域中利用低通滤波、高通滤波或带通滤波等数字图像进行处理。

图 6.1 所示为离散傅里叶变换的实验结果。图 6.1(a)和图 6.1(b)为对应于物体变形前后的两幅单曝光数字散斑图；图 6.1(c)为散斑图经过离散傅里叶变换后得到的傅里叶频谱分布；图 6.1(d)为所设计的理想带通滤波器；图 6.1(e)为带通滤波后的傅里叶频谱分布；图 6.1(f)为经过离散傅里叶反变换后得到的离面位移导数的等值条纹图。

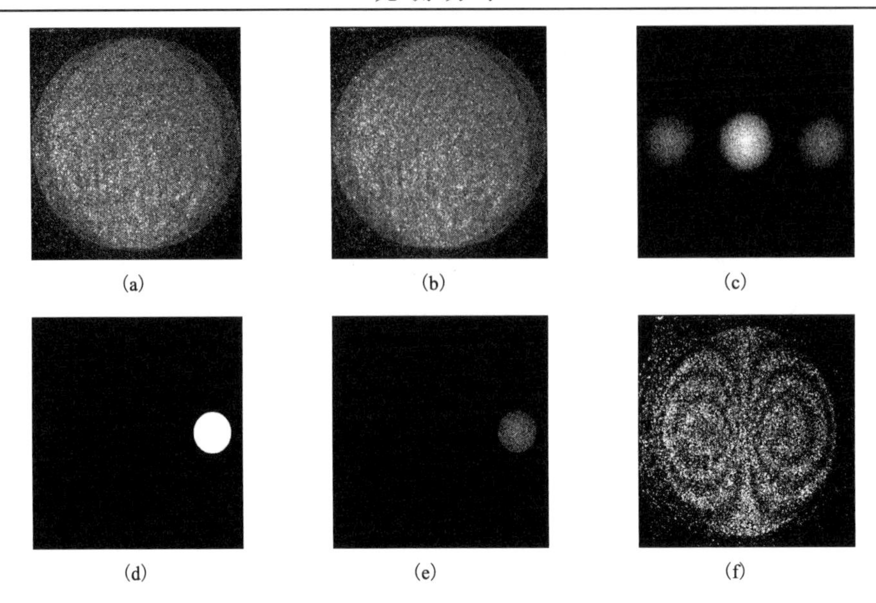

(a) (b) (c)

(d) (e) (f)

图 6.1 傅里叶变换的应用

6.1.2 离散余弦变换

尽管离散傅里叶变换在信号处理和图像处理中获得了广泛应用，但离散傅里叶变换要涉及复数运算，因此运算量较大(复数的运算量相当于实数的两倍)。为了克服傅里叶变换的上述问题，人们提出了离散余弦变换。离散余弦变换是以一组不同频率和不同幅值的余弦函数之和来近似表征一幅图像，实际上它是傅里叶变换的实数部分，因此离散余弦变换的运算量较小。

1. 离散余弦变换原理

1) 一维离散余弦变换

设 $f(m)$ 为一维实数离散序列，其中，$m=0,1,\cdots,M-1$，则一维离散余弦变换定义为

$$F(p)=C(p)\sum_{m=0}^{M-1} f(m)\cos\left[\frac{\pi(2m+1)p}{2M}\right] \quad (p=0,1,\cdots,M-1) \tag{6.15}$$

式中，p 为频域像素坐标，$C(p)=\begin{cases}1/\sqrt{M} & (p=0)\\ \sqrt{2/M} & (p=1,2,\cdots,M-1)\end{cases}$。

一维离散余弦反变换定义为

$$f(m)=C(m)\sum_{p=0}^{M-1} F(p)\cos\left[\frac{\pi(2m+1)p}{2M}\right] \quad (m=0,1,\cdots,M-1) \tag{6.16}$$

式中，$C(m)=\begin{cases}1/\sqrt{M} & (m=0)\\ \sqrt{2/M} & (m=1,2,\cdots,M-1)\end{cases}$。

显然，对于离散余弦变换，其变换和反变换具有相同的变换核。

2）二维离散余弦变换

设 $f(m,n)$ 为二维实数离散序列，其中，$m=0,1,\cdots,M-1$; $n=0,1,\cdots,N-1$，则二维离散余弦变换定义为

$$F(p,q)=C(p)C(q)\sum_{m=0}^{M-1}\sum_{n=0}^{N-1}f(m,n)\cos\left[\frac{\pi(2m+1)p}{2M}\right]\cos\left[\frac{\pi(2n+1)q}{2N}\right]\quad(p=0,1,\cdots,M-1;q=0,1,\cdots,N-1)\tag{6.17}$$

式中，$C(p)=\begin{cases}1/\sqrt{M} & (p=0)\\ \sqrt{2/M} & (p=1,2,\cdots,M-1)\end{cases}$；$C(q)=\begin{cases}1/\sqrt{N} & (q=0)\\ \sqrt{2/N} & (q=1,2,\cdots,N-1)\end{cases}$。

二维离散余弦反变换定义为

$$F(m,n)=C(m)C(n)\sum_{p=0}^{M-1}\sum_{q=0}^{N-1}f(p,q)\cos\left[\frac{\pi(2m+1)p}{2M}\right]\cos\left[\frac{\pi(2n+1)q}{2N}\right]\quad(m=0,1,\cdots,M-1;n=0,1,\cdots,N-1)\tag{6.18}$$

式中，$C(m)=\begin{cases}1/\sqrt{M} & (m=0)\\ \sqrt{2/M} & (m=1,2,\cdots,M-1)\end{cases}$；$C(n)=\begin{cases}1/\sqrt{N} & (n=0)\\ \sqrt{2/N} & (n=1,2,\cdots,N-1)\end{cases}$。

2. 离散余弦变换实验

离散余弦变换与离散傅里叶变换一样，将数字图像从空域变换到频域，然后在频域中对数字图像进行处理。离散余弦变换在光测力学中的主要应用是进行低通滤波和图像压缩。

图 6.2 所示为离散余弦变换的实验结果。图 6.2(a)和图 6.2(b)为对应于物体变形前后的两幅单曝光数字散斑图；图 6.2(c)为数字散斑图经过离散余弦变换后得到的频谱分布；图 6.2(d)为带通滤波器；图 6.2(e)为滤波后的频谱；图 6.2(f)为经过离散余弦反变换后得到的离面位移导数的等值条纹图。

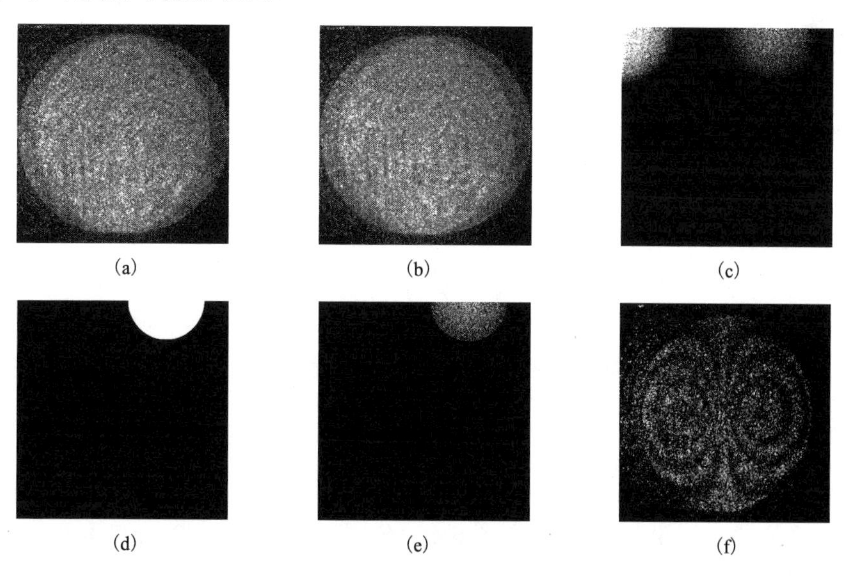

图 6.2　余弦变换的应用

6.2 低 通 滤 波

图像滤波技术分为空域滤波和频域滤波等。空域滤波包括平滑滤波和锐化滤波。频域滤波包括低通滤波、高通滤波和带通滤波。因为空域平滑滤波和频域低通滤波技术能够降低或消除条纹图高频噪声，因此这些技术已经广泛应用于光测力学条纹图降噪。

空域平滑滤波就是在空域对图像的各个像素灰度进行平滑处理。空域平滑滤波包括均值滤波、中值滤波和自适应滤波。

频域低通滤波就是在频域对图像进行低通滤波，然后再进行反变换，得到处理后的图像。对图像进行傅里叶变换和余弦变换就能得到它的频谱分布。物体的变形信息对应于频谱的低频分量，而噪声则对应于频谱的高频分量，通过滤掉高频分量就可抑制或消除噪声。频域低通滤波主要包括理想低通滤波、巴特沃思低通滤波和指数低通滤波等。

图像往往存在噪声(如图 6.3 所示)，因此在进行相位计算之前，需要选用合适的低通滤波方法进行噪声抑制或消除。图 6.3(a)、(b)和(c)为采用三步相移而得到的干涉条纹图，相移量依次为 0 、 $\pi/2$ 和 π 。

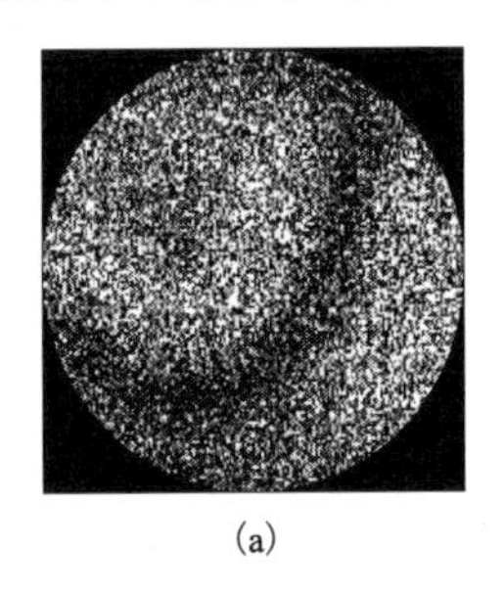
(a)

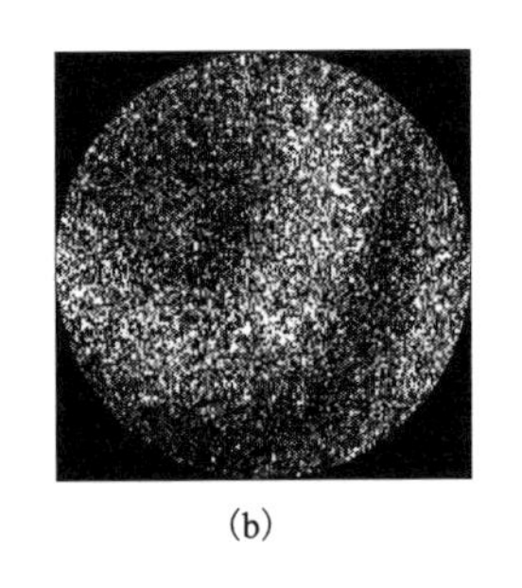
(b)

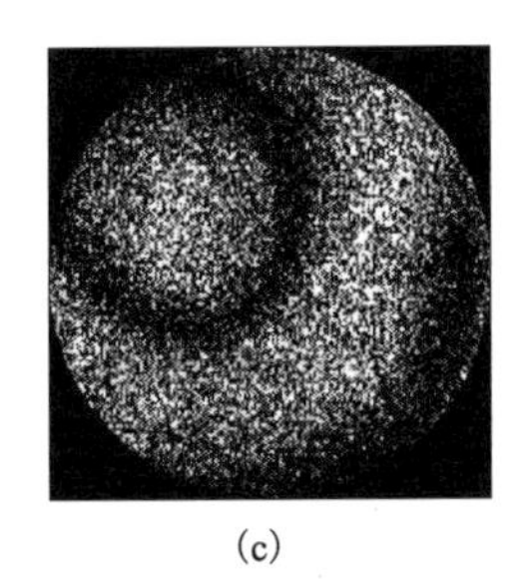
(c)

图 6.3 干涉条纹图

6.2.1 空域均值平滑滤波

1. 均值平滑滤波原理

均值平滑滤波实际上就是对像素邻域进行平均操作，将平均值作为输出像素灰度。均值平滑滤波可表示为

$$g(m,n)=\frac{1}{h}\sum_{-p,-q}^{p,q} f(m+i,n+j) \tag{6.19}$$

式中，$f(m,n)$ 和 $g(m,n)$ 分别为滤波前后像素 (m,n) 灰度；$h=(2p+1)(2q+1)$ 为滤波器尺寸。

2. 均值平滑滤波实验

图 6.4 所示为均值滤波结果。图 6.4(a)、图 6.4(b)和图 6.4(c)分别为图 6.3(a)、图 6.3(b)和图 6.3(c)的均值滤波结果；图 6.4(d)和图 6.4(e)分别为相位在 $-\pi/2\sim\pi/2$ 和 $0\sim2\pi$ 范围的包裹相位分布；图 6.4(f)为连续相位分布。

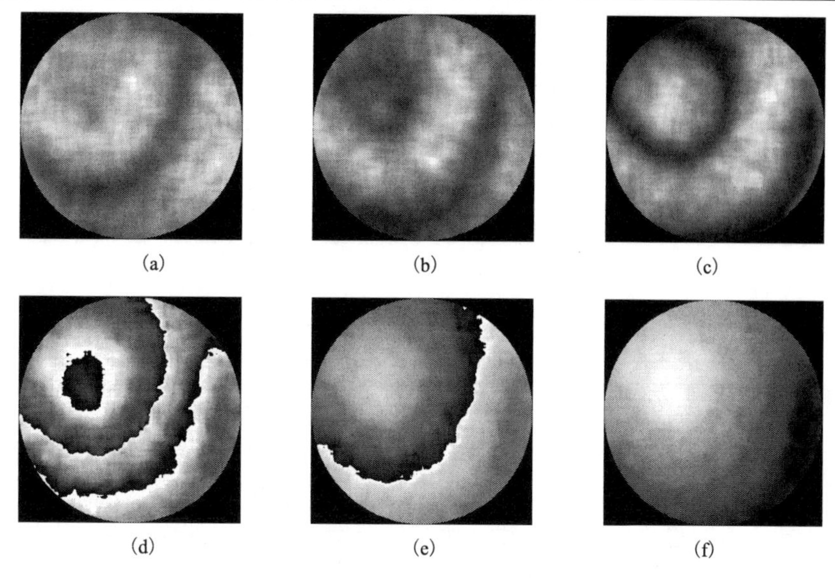

图 6.4　均值滤波结果

6.2.2　空域中值平滑滤波

1. 中值平滑滤波原理

中值滤波是非线性低通滤波方法，它可以有效保护图像边缘，同时可以去除噪声。与均值滤波不同，中值滤波是将邻域中的像素按灰度大小排序，取中间值为输出像素。中值平滑滤波可表示为

$$g(m,n)=\mathrm{MF}\{f(m-p,n-q),\cdots,f(m-p,n+q),\cdots,f(m+p,n-q),\cdots,f(m+p,n+q)\} \tag{6.20}$$

式中，$f(m,n)$ 和 $g(m,n)$ 分别为滤波前后像素 (m,n) 灰度；$(2p+1)(2q+1)$ 为滤波器尺寸；$\mathrm{MF}\{\cdots\}$ 表示中值滤波。

2. 中值平滑滤波实验

图 6.5 所示为中值滤波结果。图 6.5(a)、图 6.5(b)和图 6.5(c)分别为图 6.3(a)、图 6.3(b)和图 6.3(c)的中值滤波结果；图 6.5(d)和图 6.5(e)分别为相位在 $-\pi/2\sim\pi/2$ 和 $0\sim2\pi$ 范围的包裹相位分布；图 6.5(f)为连续相位分布。

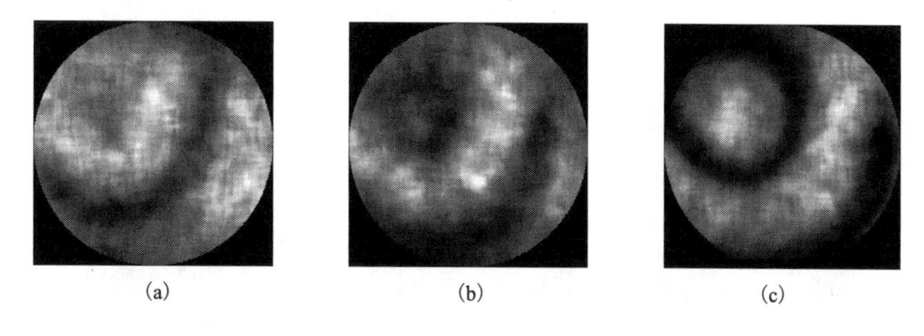

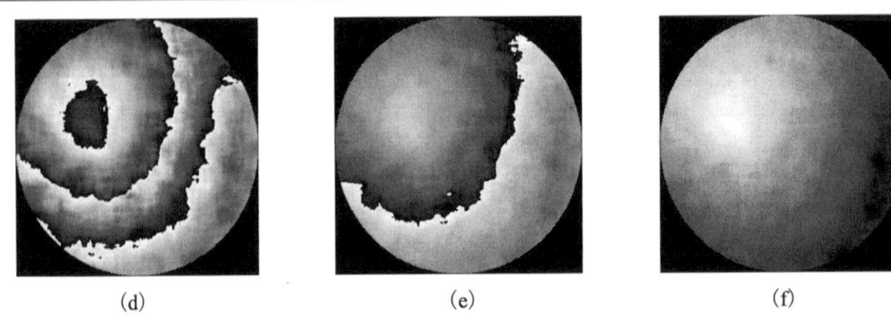

(d)　(e)　(f)

图 6.5　中值滤波结果

6.2.3　空域自适应平滑滤波

1. 自适应平滑滤波原理

自适应滤波先对图像进行均值滤波，然后根据局部方差和图像噪声对均值进行修正。自适应平滑滤波可表示为

$$g(m,n)=f(m,n)-\sigma^2\frac{f(m,n)-\dfrac{1}{h}\sum\limits_{-p,-q}^{p,q}f(m+i,n+j)}{\dfrac{1}{h}\sum\limits_{-p,-q}^{p,q}[f(m+i,n+j)]^2-\left[\dfrac{1}{h}\sum\limits_{-p,-q}^{p,q}f(m+i,n+j)\right]^2} \tag{6.21}$$

式中，$f(m,n)$ 和 $g(m,n)$ 分别为滤波前后像素 (m,n) 灰度；$h=(2p+1)(2q+1)$ 为滤波器尺寸，σ^2 为图像噪声。

2. 自适应平滑滤波实验

图 6.6 所示为自适应滤波结果。图 6.6(a)、图 6.6(b)和图 6.6(c)分别为图 6.3(a)、图 6.3(b)和图 6.3(c)的自适应滤波结果；图 6.6(d)和图 6.6(e)分别为相位在 $-\pi/2\sim\pi/2$ 和 $0\sim2\pi$ 范围的包裹相位分布；图 6.6(f)为连续相位分布。

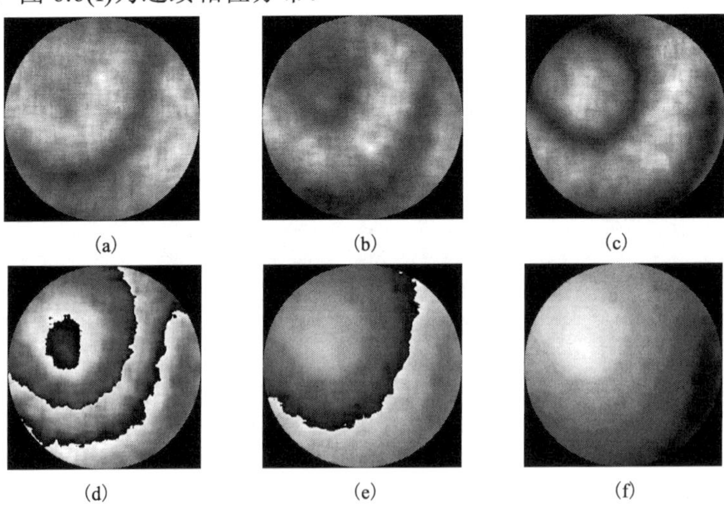

(a)　(b)　(c)

(d)　(e)　(f)

图 6.6　自适应滤波结果

6.2.4　频域理想低通滤波

1. 理想低通滤波原理

理想低通滤波器定义为

$$H(p,q)=\begin{cases}1 & (D(p,q)\leqslant D_0)\\0 & (D(p,q)>D_0)\end{cases} \tag{6.22}$$

式中，$H(p,q)$ 为传递函数；D_0 为截止频率；$D(p,q)=\sqrt{p^2+q^2}$ 。

2. 理想低通滤波实验

图 6.7 和图 6.8 所示为理想低通滤波结果。图 6.7(a)、图 6.7(b)和图 6.7(c)分别为图 6.3(a)、图 6.3(b)和图 6.3(c)的傅里叶变换理想低通滤波结果。图 6.8(a)、图 6.8(b)和图 6.8(c)分别为图 6.3(a)、图 6.3(b)和图 6.3(c)的余弦变换理想低通滤波结果。图 6.7 和图 6.8 中的(d)和(e)分别为相位在 $-\pi/2\sim\pi/2$ 和 $0\sim2\pi$ 范围内的包裹相位分布，图 6.7(f)和图 6.8(f)分别为连续相位分布。

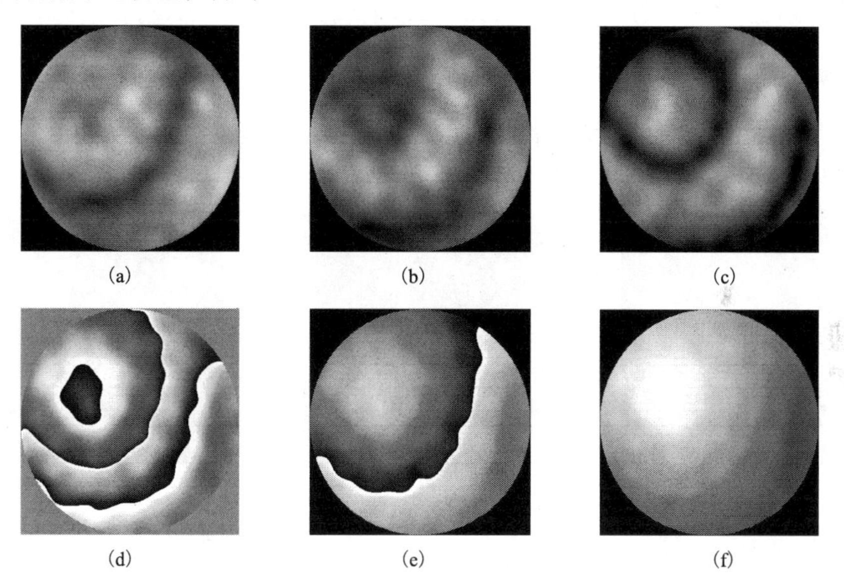

图 6.7　傅里叶变换理想低通滤波结果

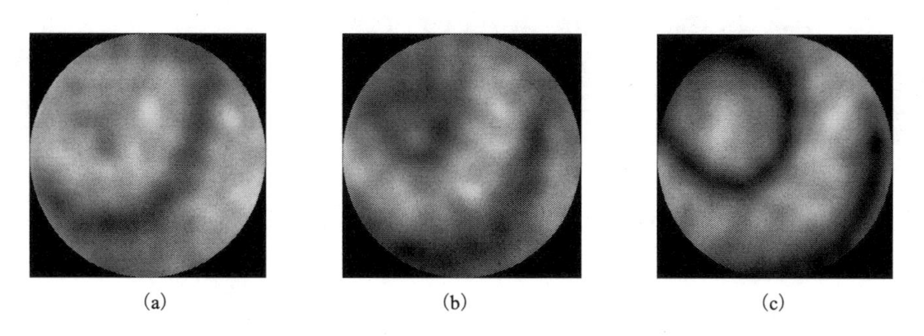

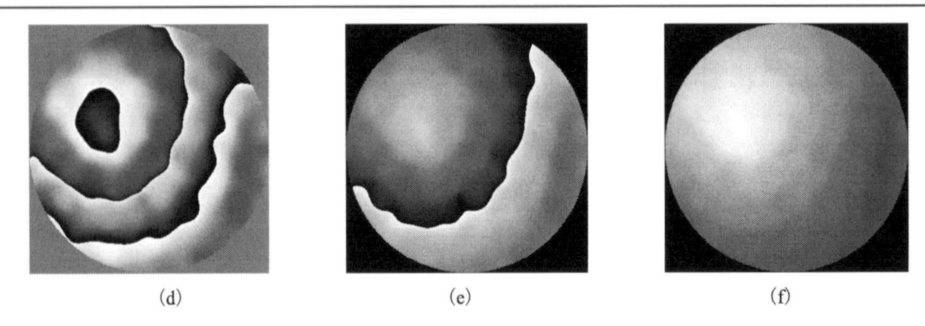

(d)　(e)　(f)

图 6.8　余弦变换理想低通滤波结果

6.2.5　频域巴特沃思低通滤波

1. 巴特沃思低通滤波原理

巴特沃思低通滤波器定义为

$$H(p,q)=\frac{1}{1+\left[D(p,q)/D_0\right]^{2n}} \tag{6.23}$$

式中，D_0 为截止频率；n 为滤波器的阶数。当 $D(p,q)=D_0$ 时，$H(p,q)=0.5$。

2. 巴特沃思低通滤波实验

图 6.9 和图 6.10 所示分别为巴特沃思低通滤波结果，其中，图 6.9 采用傅里叶变换，图 6.10 采用余弦变换。

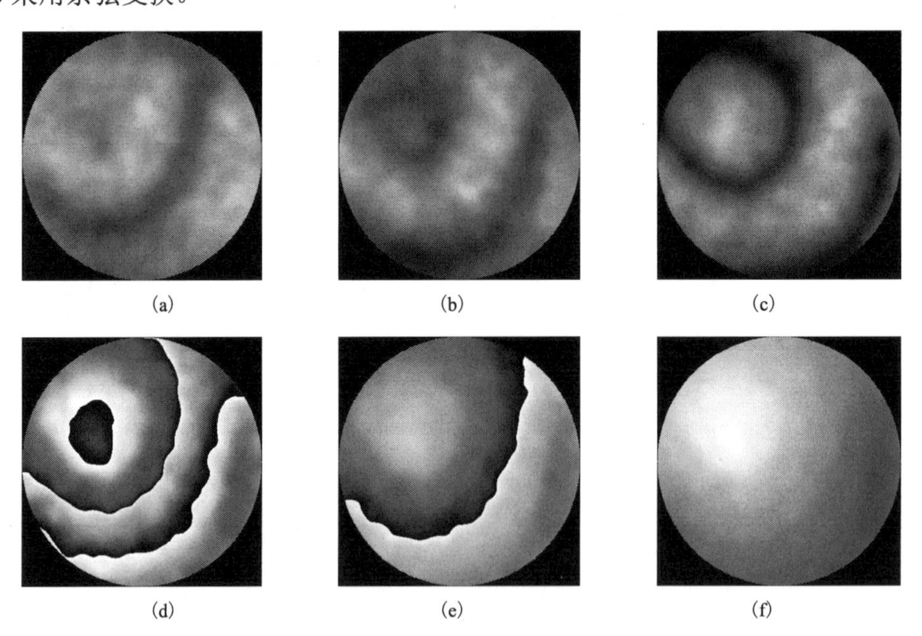

(a)　(b)　(c)

(d)　(e)　(f)

图 6.9　傅里叶变换巴特沃思低通滤波结果

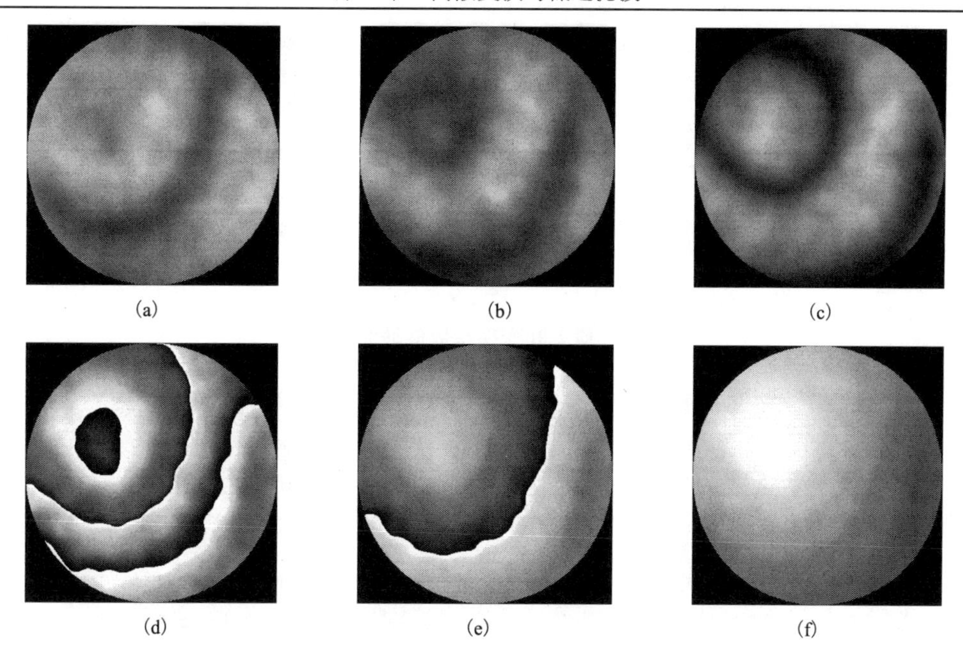

(a)　(b)　(c)

(d)　(e)　(f)

图 6.10　余弦变换巴特沃思低通滤波结果

6.2.6　频域指数低通滤波

1. 指数低通滤波原理

指数低通滤波器定义为

$$H(p,q)=\exp\left\{-\left[\frac{D(p,q)}{D_0}\right]^n\right\} \tag{6.24}$$

式中，D_0 为截止频率；n 为滤波器衰减的系数。当 $D(p,q)=D_0$ 时，$H(p,q)\approx 1/2.7$。

2. 指数低通滤波实验

图 6.11 和图 6.12 所示分别为指数低通滤波结果，其中，图 6.11 采用傅里叶变换，图 6.12 采用余弦变换。

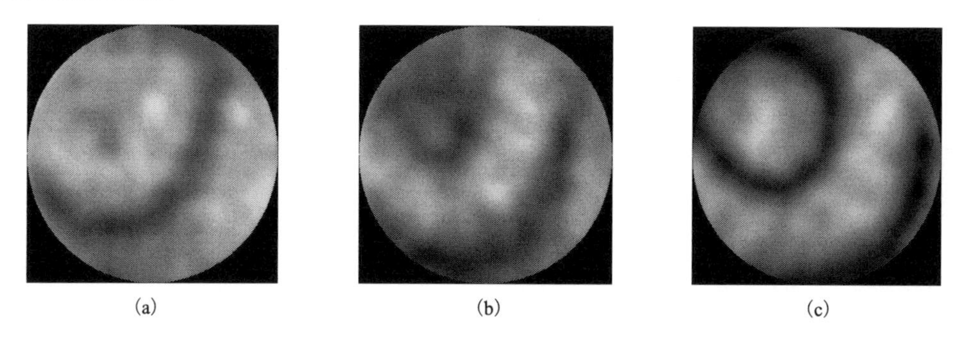

(a)　(b)　(c)

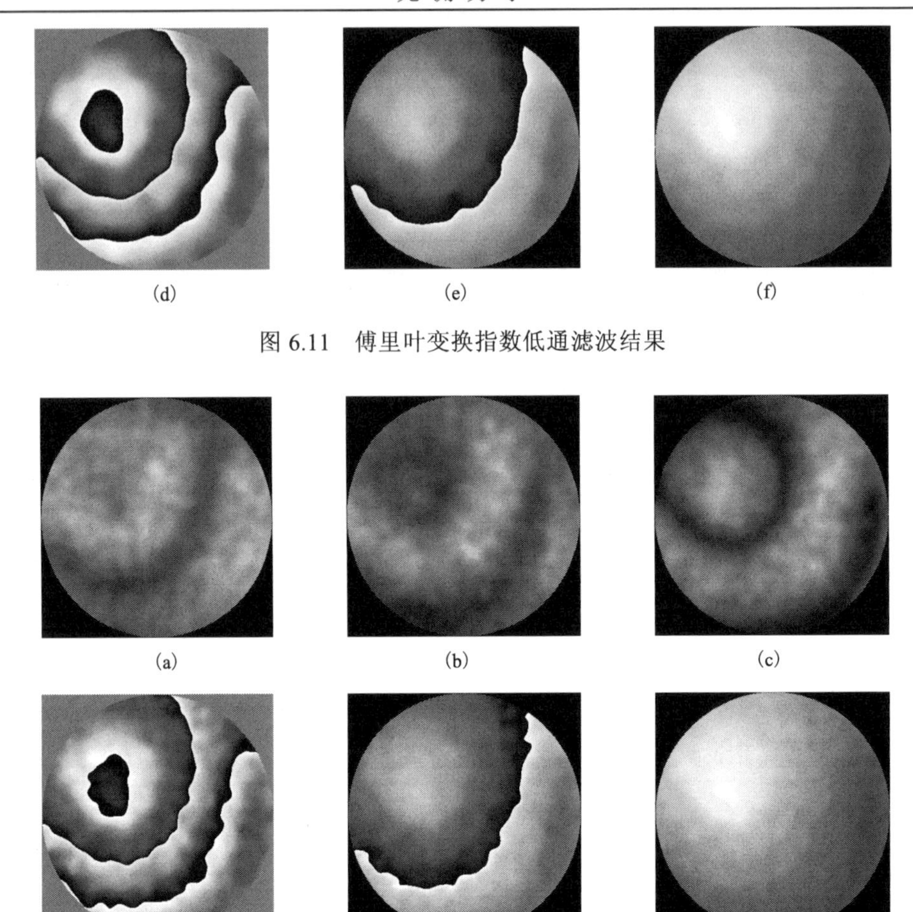

图 6.11　傅里叶变换指数低通滤波结果

图 6.12　余弦变换指数低通滤波结果

第7章　数字全息照相与数字全息干涉

全息照相记录的是物体光波和参考光波发生干涉而形成的干涉条纹。由于干涉条纹的空间频率通常很高，因此要求记录介质具有很高的分辨率。自从全息技术提出以来，记录介质主要采用具有很高分辨率的照相底片，但由于其感光灵敏度低，所需曝光时间长，因而对记录系统的稳定性具有较高的要求。另外，照相底片记录全息图后，需要进行显影和定影等冲洗处理。为了克服全息照相的上述缺点，人们提出了数字全息照相，即采用光敏电子器件(如 CCD 或 CMOS)代替传统记录介质来完成全息记录。采用光敏电子器件作为记录介质时无须显影和定影等冲洗过程，因而简化了记录过程。另外，可以采用数字方法，模拟光波衍射来再现物体光波，因而省去了光学再现装置。

数字全息虽然具有诸多优点，但目前数字全息也存在一些不足。同传统记录介质(如照相底片)比较，光敏电子器件的空间分辨率还比较低，光敏面尺寸还比较小，使得现阶段数字全息再现像的分辨率不高。然而，随着计算机科学和光敏电子器件的快速发展，这些不足将逐步被克服，数字全息技术将得到更大的发展和更广泛的应用。

7.1　数字全息照相

光学全息的基础理论与实验技术同样适用于数字全息。但由于目前记录数字全息图的光敏电子器件的光敏面比较小，空间分辨率比较低，因此数字全息只能在有限距离内记录和再现较小物体。数字全息需要满足光学全息的所有要求，还需要在记录过程中满足采样定理。

目前，数字全息记录常用的光敏电子器件是 CCD。由于 CCD 作为记录介质具有极高的感光灵敏度和较宽的波长响应范围，因而在全息记录方面具有极大的优势。另外，由于 CCD 价格比较便宜，因此目前 CCD 已广泛应用于数字全息技术。

7.1.1　数字全息记录

数字全息记录系统与传统全息记录系统基本相同，主要不同是用 CCD 取代照相底片作为记录介质。图 7.1 所示为离轴数字全息记录系统。来自激光器的光波经分光镜分束后变成两束光波，其中一束为物体光波，该光波经反射镜反射并经扩束镜扩束后照明物体，然后经物体漫反射后再垂直照射 CCD 靶面；另一束为参考光波，该光波经反射镜反射并经扩束镜扩束后直接照射 CCD 靶面。物体光波和参考光波在 CCD 靶面由于相干叠加而形成菲涅耳全息图。

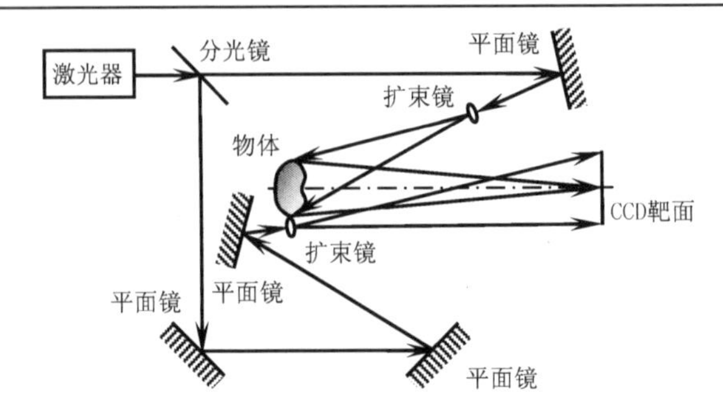

图 7.1　离轴数字全息记录系统

设物面上的物体光波复振幅为 $O(x_o, y_o)$，则在菲涅耳衍射区，CCD 靶面处物体光波复振幅 $O(x, y)$ 可表示为

$$
\begin{aligned}
O(x,y)=&\frac{\exp(\mathrm{i}kz_o)}{\mathrm{i}\lambda z_o}\exp\left[\mathrm{i}\frac{\pi}{\lambda z_o}(x^2+y^2)\right]\int_{-\infty}^{\infty}\int_{-\infty}^{\infty}O(x_o,y_o)\\
&\bullet\exp\left[\mathrm{i}\frac{\pi}{\lambda z_o}(x_o^2+y_o^2)\right]\exp\left[-\mathrm{i}\frac{2\pi}{\lambda z_o}(xx_o+yy_o)\right]\mathrm{d}x_o\mathrm{d}y_o
\end{aligned}
\tag{7.1}
$$

式中，$\mathrm{i}=\sqrt{-1}$ 是虚数单位；λ 是波长；$k=2\pi/\lambda$ 是波数；z_o 是物面到 CCD 靶面的距离。

设照射 CCD 靶面的参考光波复振幅为 $R(x, y)$，则由于物体光波和参考光波的干涉而在 CCD 靶面产生的强度分布为

$$
I(x,y)=|O(x,y)|^2+|R(x,y)|^2+O(x,y)R^*(x,y)+O^*(x,y)R(x,y) \tag{7.2}
$$

式中，$*$表示复共轭，；$|O(x,y)|^2$ 和 $|R(x,y)|^2$ 分别是物体光波和参考光波的强度分布；$O(x,y)R^*(x,y)$ 和 $O^*(x,y)R(x,y)$ 都是干涉项，分别包含物体光波的振幅和相位信息。

通过 CCD 记录的全息图称为数字全息图。设 CCD 的有效像素为 $M\times N$，相邻像素在 x 和 y 方向的中心间距分别为 Δx 和 Δy，则 CCD 记录的数字全息图的离散强度分布为

$$
I(m,n)=I(x,y)S(m,n)=\sum_{m=0}^{M-1}\sum_{n=0}^{N-1}I(x,y)\delta(x-m\Delta x,y-n\Delta y) \tag{7.3}
$$

式中

$$
S(m,n)=\sum_{m=0}^{M-1}\sum_{n=0}^{N-1}\delta(x-m\Delta x,y-n\Delta y) \tag{7.4}
$$

为采样函数。

7.1.2　数字全息再现

在数字全息中，采用数值方法，模拟光波衍射，再现物体光波。设再现光波复振幅为 $C(x, y)$，则透过全息图的光波复振幅可表示为

$$
A(m,n)=C(x,y)I(m,n)=\sum_{m=0}^{M-1}\sum_{n=0}^{N-1}C(x,y)I(x,y)\delta(x-m\Delta x,y-n\Delta y) \tag{7.5}
$$

在菲涅耳衍射区，距离全息图为 z_r 处的衍射光波复振幅分布为

$$A(p,q)=\frac{\exp(\mathrm{i}kz_{\mathrm{r}})}{\mathrm{i}\lambda z_{\mathrm{r}}}\exp\left[\mathrm{i}\frac{\pi}{\lambda z_{\mathrm{r}}}(p^2\Delta x_{\mathrm{r}}^{\ 2}+q^2\Delta y_{\mathrm{r}}^{\ 2})\right]$$
$$\times\sum_{m=0}^{M-1}\sum_{n=0}^{N-1}A(m,n)\exp\left[\mathrm{i}\frac{\pi}{\lambda z_{\mathrm{r}}}(m^2\Delta x^2+n^2\Delta y^2)\right]\exp\left[-\mathrm{i}2\pi(\frac{mp}{M}+\frac{nq}{N})\right] \tag{7.6}$$
$$(p=0,1,\cdots,M-1;q=0,1,\cdots,N-1)$$

式中，Δx_{r} 和 Δy_{r} 为距离全息图为 z_{r} 处观察面上相邻像素分别在 x 和 y 方向的中心间距，可表示为

$$\Delta x_{\mathrm{r}}=\frac{\lambda z_{\mathrm{r}}}{M\Delta x},\quad \Delta y_{\mathrm{r}}=\frac{\lambda z_{\mathrm{r}}}{N\Delta y} \tag{7.7}$$

把式(7.7)代入式(7.6)，得

$$A(p,q)=\frac{\exp(\mathrm{i}kz_{\mathrm{r}})}{\mathrm{i}\lambda z_{\mathrm{r}}}\exp\left[\mathrm{i}\pi\lambda z_{\mathrm{r}}\left(\frac{p^2}{M^2\Delta x^2}+\frac{q^2}{N^2\Delta y^2}\right)\right]$$
$$\times\sum_{m=0}^{M-1}\sum_{n=0}^{N-1}A(m,n)\exp\left[\mathrm{i}\frac{\pi}{\lambda z_{\mathrm{r}}}(m^2\Delta x^2+n^2\Delta y^2)\right]\exp\left[-\mathrm{i}2\pi\left(\frac{mp}{M}+\frac{nq}{N}\right)\right] \tag{7.8}$$
$$(p=0,1,\cdots,M-1;q=0,1,\cdots,N-1)$$

因此，强度分布为

$$I(p,q)=|A(p,q)|^2\quad (p=0,1,\cdots,M-1;q=0,1,\cdots,N-1) \tag{7.9}$$

相位分布为

$$\varphi(p,q)=\arctan\frac{\mathrm{Im}\{A(p,q)\}}{\mathrm{Re}\{A(p,q)\}}\quad (p=0,1,\cdots,M-1;q=0,1,\cdots,N-1) \tag{7.10}$$

式中，$\mathrm{Re}\{\cdot\}$ 和 $\mathrm{Im}\{\cdot\}$ 分别表示取实部和虚部。对于粗糙表面，相位 $\varphi(p,q)$ 随机变化，而在数字全息照相中感兴趣的是强度 $I(p,q)$。

在数字全息照相中，经常也会采用如图 7.2 所示的同轴数字全息记录系统。

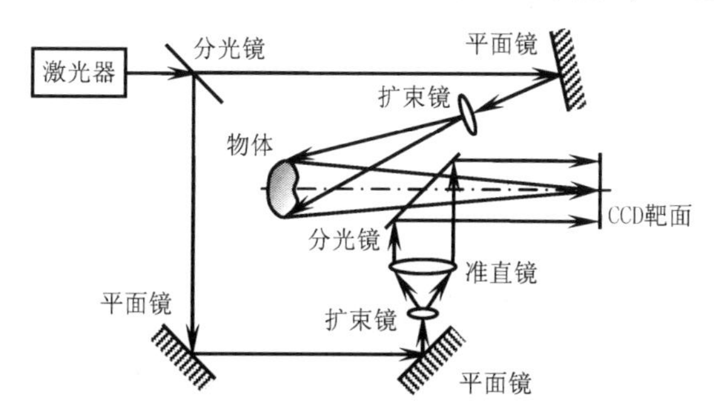

图 7.2　同轴数字全息记录系统

7.1.3　数字全息实验

实验一　数字全息图如图 7.3(a)所示，再现像如图 7.3(b)所示。

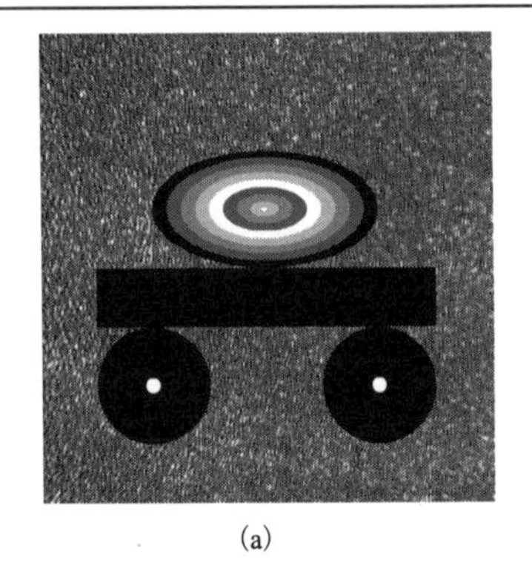

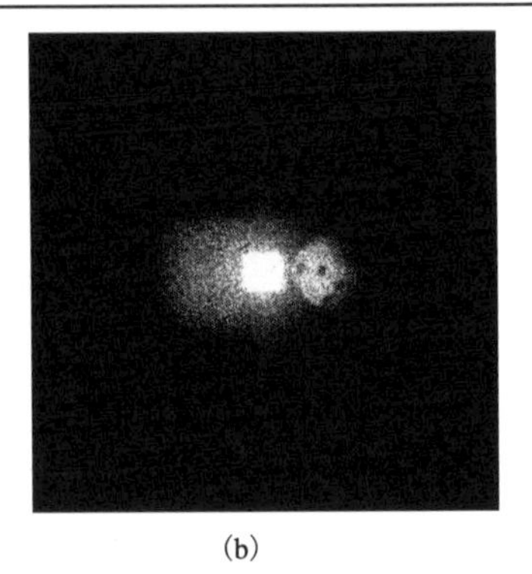

(a)　　(b)

图 7.3　实验一所得再现结果

实验二　数字全息图如图 7.4(a)所示，再现像如图 7.4(b)所示。

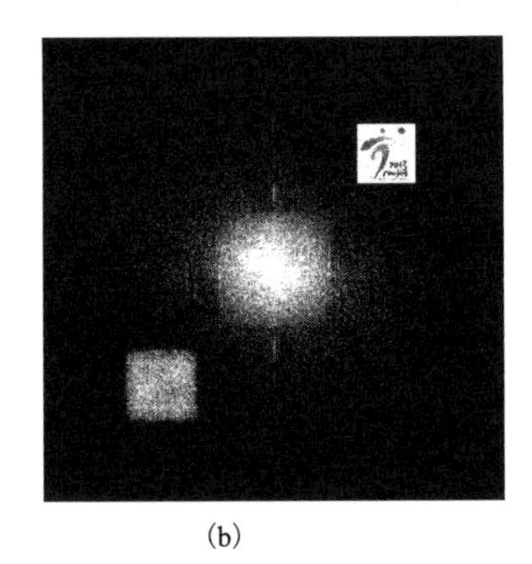

(a)　　(b)

图 7.4　实验二所得再现结果

7.2　数字全息干涉

1. 数字全息干涉原理

与传统全息干涉一样，数字全息干涉也可用于物体的变形测量。设物体变形前后的复振幅分布分别由 $A_1(p,q)$ 和 $A_2(p,q)$ 表示，则物体变形前后的相位分布可分别表示为

$$\varphi_1(p,q)=\arctan\frac{\mathrm{Im}\{A_1(p,q)\}}{\mathrm{Re}\{A_1(p,q)\}},\quad \varphi_2(p,q)=\arctan\frac{\mathrm{Im}\{A_2(p,q)\}}{\mathrm{Re}\{A_2(p,q)\}} \tag{7.11}$$
$$(p=0,1,\cdots,M-1;q=0,1,\cdots,N-1)$$

物体变形前后的相位分布都是随机变量，但其差值将不再是随机变量，表示因物体受载而引起的相位变化，仅与物体变形有关。利用式(7.11)，则由物体变形而引起的相位变化可表示为

$$\delta(p,q)=\varphi_2(p,q)-\varphi_1(p,q)=\arctan\frac{\mathrm{Im}\{A_2(p,q)\}}{\mathrm{Re}\{A_2(p,q)\}}-\arctan\frac{\mathrm{Im}\{A_1(p,q)\}}{\mathrm{Re}\{A_1(p,q)\}} \tag{7.12}$$
$$(p=0,1,\cdots,M-1;q=0,1,\cdots,N-1)$$

式中，$\delta(p,q)$是包裹相位。通过相位展开，即可获得连续相位分布。

2. 数字全息干涉实验

图 7.5 是采用数字全息干涉得到的相位分布。图 7.5(a)和图 7.5(b)分别为物体变形前后的相位分布，图 7.5(c)为图 7.5(a)和图 7.5(b)相减后而得到的与物体变形有关的包裹相位分布；图 7.5(d)为图 7.5(c)进行相位展开而得到的连续相位分布。

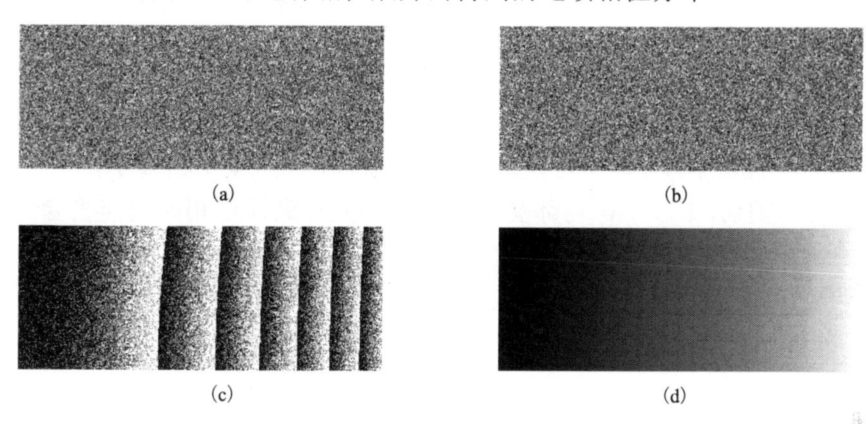

图 7.5　相位分布

第 8 章　数字散斑干涉与数字散斑剪切干涉

数字散斑(剪切)干涉的基本原理与传统散斑(剪切)干涉相同。传统方法由于采用照相底片记录散斑图，因此需要进行显影和定影等冲洗处理。另外，在传统方法中，是将两次曝光记录相加，因此需要进行滤波，以显现干涉条纹。数字方法由于采用 CCD 记录数字散斑图，因此不需要进行显影和定影；另外，通过 CCD 记录的数字散斑图可以相加存储，也可以分开存储，因此数字方法除了可以采用相加方法，还可以采用相减方法。

在数字散斑(剪切)干涉中，物体变形前后的曝光记录通过相减就能去除背景强度，因此数字散斑(剪切)干涉主要采用相减方法。采用相减方法不需要进行滤波即可显现干涉条纹。由于变形前后的曝光记录被独立地进行处理，因此相移干涉技术能够很方便地应用于数字散斑(剪切)干涉。

8.1　数字散斑干涉

8.1.1　面内位移测量

测量面内位移的数字散斑干涉系统如图 8.1 所示。两束准直光波对称照射物面，两束光波与物面法线的夹角均为 θ。散射光波在 CCD 靶面相互干涉。

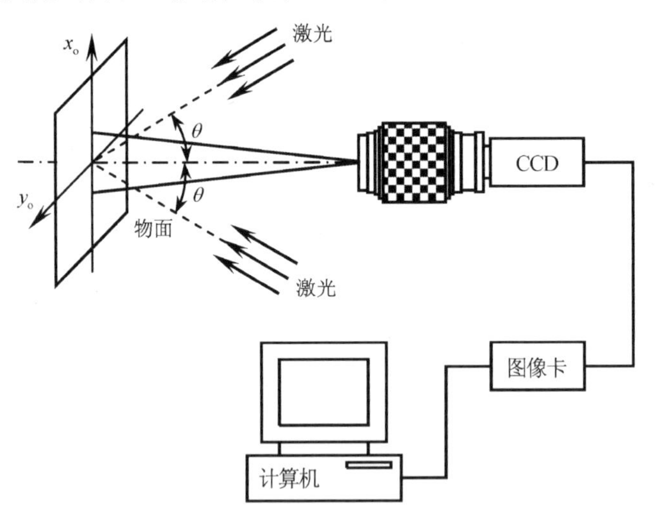

图 8.1　面内位移数字散斑干涉系统

1. 条纹形成

物体变形前 CCD 记录的强度分布可表示为

$$I_1 = I_{o1} + I_{o2} + 2\sqrt{I_{o1}I_{o2}}\cos\varphi \tag{8.1}$$

式中，I_{o1} 和 I_{o2} 分别为两束入射光波的强度分布；φ 为两束入射光波的相位差。

物体变形后 CCD 记录的强度分布为

$$I_2 = I_{o1} + I_{o2} + 2\sqrt{I_{o1}I_{o2}}\cos(\varphi+\delta) \tag{8.2}$$

式中，δ 为因物体变形而引起的两束入射光波的相对相位变化，可表示为

$$\delta = 2ku_o\sin\theta \tag{8.3}$$

式中，k 为波数；u_o 为物面沿 x 方向的面内位移分量。

物体变形前后所记录的强度相减所得差的平方可表示为

$$E = (I_2 - I_1)^2 = 8I_{o1}I_{o2}\sin^2\left(\varphi + \frac{1}{2}\delta\right)(1-\cos\delta) \tag{8.4}$$

式中，正弦项为高频成分，对应于散斑噪声；余弦项为低频成分，对应于物体变形。因此当满足条件

$$\delta = 2n\pi \quad (n = 0, \pm 1, \pm 2, \cdots) \tag{8.5}$$

时，条纹亮度将达到最小，即暗纹将产生于

$$u_o = \frac{n\pi}{k\sin\theta} = \frac{n\lambda}{2\sin\theta} \quad (n = 0, \pm 1, \pm 2, \cdots) \tag{8.6}$$

当满足条件

$$\delta = (2n+1)\pi \quad (n = 0,\ \pm 1, \pm 2, \cdots) \tag{8.7}$$

时，条纹亮度将达到最大，即亮纹将产生于

$$u_o = \frac{(2n+1)\pi}{2k\sin\theta} = \frac{(2n+1)\lambda}{4\sin\theta} \quad (n = 0, \pm 1, \pm 2, \cdots) \tag{8.8}$$

2. 相位分析

物体变形前首先记录一幅数字散斑图，物体变形后再记录相移量依次为 0 、$\pi/2$ 、π 和 $3\pi/2$ 的 4 幅数字散斑图。采用相减模式，物体变形后的 4 幅散斑图与物体变形前的散斑图分别进行相减并平方，得

$$\begin{aligned}
E_1 &= 8I_{o1}I_{o2}\sin^2\left(\varphi + \frac{1}{2}\delta\right)(1-\cos\delta) \\
E_2 &= 8I_{o1}I_{o2}\sin^2\left(\varphi + \frac{1}{2}\delta + \frac{1}{4}\pi\right)(1+\sin\delta) \\
E_3 &= 8I_{o1}I_{o2}\sin^2\left(\varphi + \frac{1}{2}\delta + \frac{1}{2}\pi\right)(1+\cos\delta) \\
E_4 &= 8I_{o1}I_{o2}\sin^2\left(\varphi + \frac{1}{2}\delta + \frac{3}{4}\pi\right)(1-\sin\delta)
\end{aligned} \tag{8.9}$$

式中，正弦项对应高频噪声。通过低通滤波可以平均掉正弦项，由此可得

$$
\begin{aligned}
<E_1> &= 4<I_{o1}><I_{o2}>(1-\cos\delta) \\
<E_2> &= 4<I_{o1}><I_{o2}>(1+\sin\delta) \\
<E_3> &= 4<I_{o1}><I_{o2}>(1+\cos\delta) \\
<E_4> &= 4<I_{o1}><I_{o2}>(1-\sin\delta)
\end{aligned}
\tag{8.10}
$$

式中，$<\cdot>$表示系综平均。联立求解，可得物体变形相位分布为

$$
\delta = \arctan\frac{S}{C} = \arctan\frac{<E_2>-<E_4>}{<E_3>-<E_4>} \tag{8.11}
$$

式中，δ 是位于 $-\pi/2 \sim \pi/2$ 范围内的包裹相位；$S=<E_2>-<E_4>$；$C=<E_3>-<E_1>$。上述包裹相位通过如下变换可扩展到 $0\sim 2\pi$ 范围：

$$
\delta = \begin{cases}
\delta & (S \geqslant 0,\ C > 0) \\
\dfrac{1}{2}\pi & (S > 0,\ C = 0) \\
\delta + \pi & (C < 0) \\
\dfrac{3}{2}\pi & (S < 0,\ C = 0) \\
\delta + 2\pi & (S < 0,\ C > 0)
\end{cases}
\tag{8.12}
$$

由式(8.12)得到的相位分布仍然是包裹相位，因此要得到连续相位分布则需要对上述包裹相位进行相位展开。如果相邻像素之间的相位差达到或超过 π，则通过增加或减少 $2n\pi$ 的相位，可消除相位的不连续性，因此展开相位可表示为

$$
\delta_{\mathrm{u}} = \delta + 2n\pi \quad (n = \pm 1, \pm 2, \pm 3, \cdots) \tag{8.13}
$$

式中，δ_{u} 表示连续相位分布。得到连续相位分布后，面内位移分布可表示为

$$
u_{\mathrm{o}} = \frac{\delta_{\mathrm{u}}}{2k\sin\theta} = \frac{\lambda\delta_{\mathrm{u}}}{4\pi\sin\theta} \tag{8.14}
$$

3. 实验验证

图 8.2 所示为在面内位移测量中获得的相移量分别为 0、$\pi/2$、π 和 $3\pi/2$ 的 4 幅等值条纹图。

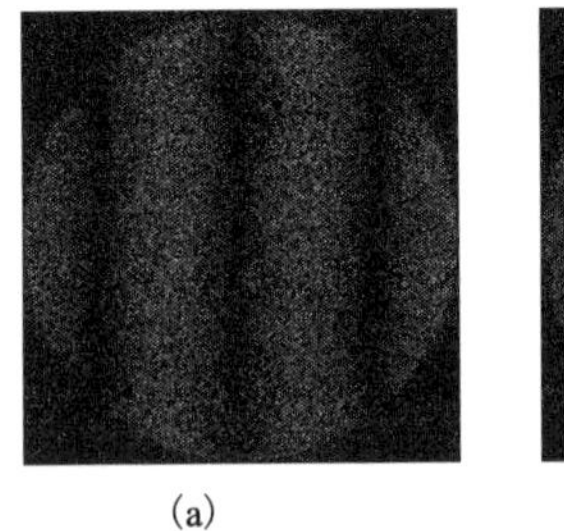
(a)

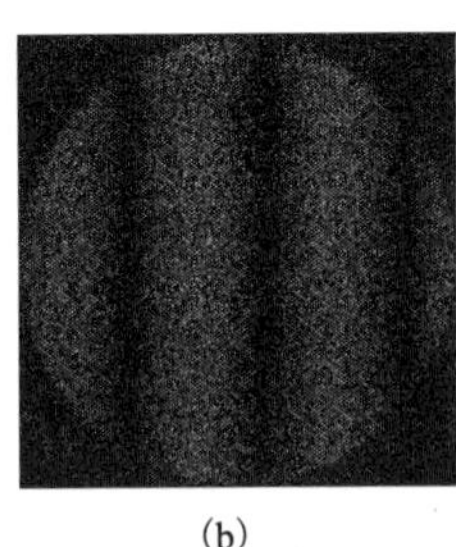
(b)

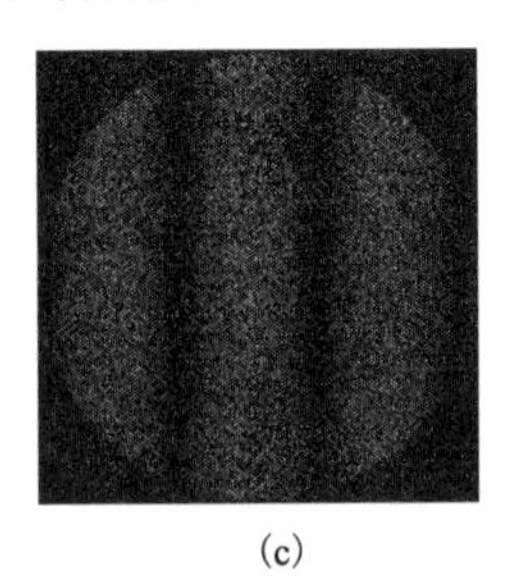
(c)

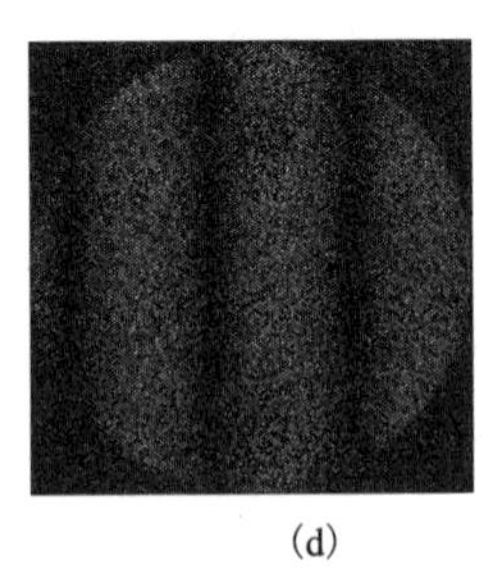
(d)

图 8.2　面内位移等值条纹

图 8.3 所示为面内位移相位图，其中图 8.3(a)是 $0\sim 2\pi$ 范围内的包裹相位分布；图 8.3(b)是连续相位分布。

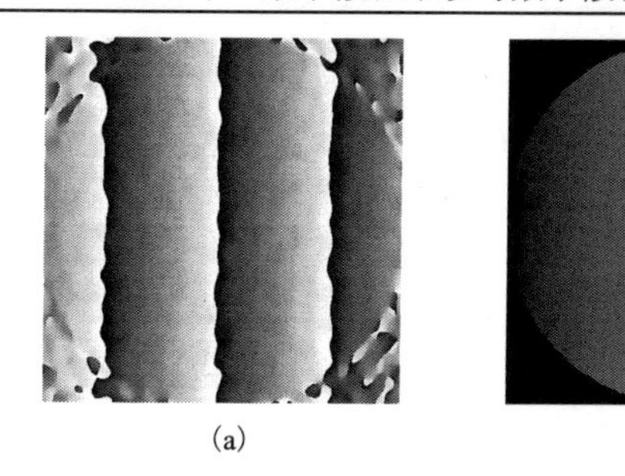

(a)　　(b)

图 8.3　面内位移相位图

8.1.2　离面位移测量

图 8.4 所示为离面位移测量系统。

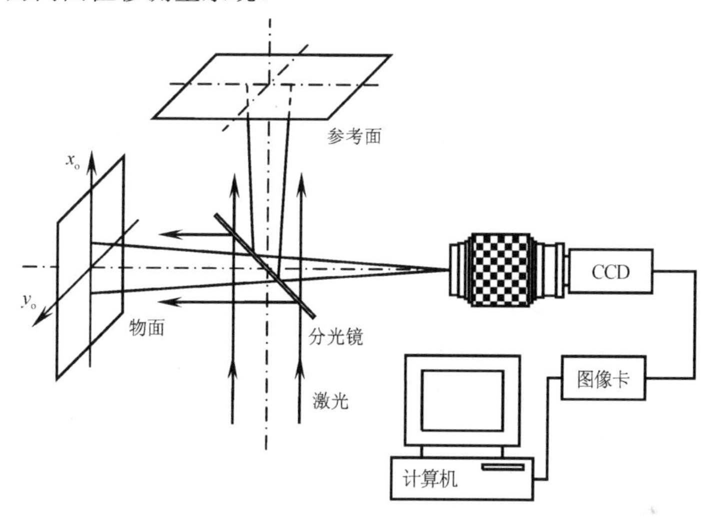

图 8.4　离面位移测量系统

1. 条纹形成

物体变形前记录的强度分布为

$$I_1 = I_o + I_r + 2\sqrt{I_o I_r}\cos\varphi \tag{8.15}$$

式中，I_o 和 I_r 分别为对应于物体光波和参考光波的强度分布；φ 为两光波之间的相位差。

物体变形后记录的强度分布为

$$I_2 = I_o + I_r + 2\sqrt{I_o I_r}\cos(\varphi + \delta) \tag{8.16}$$

式中

$$\delta = 2kw_o \tag{8.17}$$

式中，w_o 为离面位移分量。

采用相减方法，两幅数字散斑图相减所得差的平方可表示为

$$E = (I_2 - I_1)^2 = 8I_o I_r \sin^2\left(\varphi + \frac{1}{2}\delta\right)(1 - \cos\delta) \tag{8.18}$$

因此当满足条件

$$\delta = 2n\pi \quad (n = 0, \pm1, \pm2, \cdots) \tag{8.19}$$

时，条纹亮度将最小，即暗纹产生于

$$w_o = \frac{n\pi}{k} = \frac{n\lambda}{2} \quad (n = 0, \pm1, \pm2, \cdots) \tag{8.20}$$

当满足条件

$$\delta = (2n+1)\pi \quad (n = 0, \pm1, \pm2, \cdots) \tag{8.21}$$

时，条纹亮度将最大，即亮纹产生于

$$w_o = \frac{(2n+1)\pi}{2k} = \frac{(2n+1)\lambda}{4} \quad (n = 0, \pm1, \pm2, \cdots) \tag{8.22}$$

2. 相位分析

物体变形前首先记录一幅数字散斑图，物体变形后再记录相移量分别为 0 、π/2 、π 和 3π/2 的 4 幅数字散斑图。物体变形后的 4 幅数字散斑图与物体变形前的数字散斑图进行相减并平方后，即可得到相移量分别为 0 、π/2 、π 和 3π/2 的 4 幅离面位移等值条纹图。这些条纹图经过低通滤波，系综平均可分别表示为

$$\begin{aligned}
<E_1> &= 4<I_o><I_r>(1-\cos\delta) \\
<E_2> &= 4<I_o><I_r>(1+\sin\delta) \\
<E_3> &= 4<I_o><I_r>(1+\cos\delta) \\
<E_4> &= 4<I_o><I_r>(1-\sin\delta)
\end{aligned} \tag{8.23}$$

联立求解，可得物体变形相位分布为

$$\delta = \arctan\frac{<E_2> - <E_4>}{<E_3> - <E_1>} \tag{8.24}$$

式中，δ 是位于 $-\pi/2 \sim \pi/2$ 范围内的包裹相位。通过相位扩展和相位展开，可得到连续相位分布。根据连续相位分布，离面位移分布可表示为

$$w_o = \frac{\delta_u}{2k} = \frac{\lambda\delta_u}{4\pi} \tag{8.25}$$

3. 实验验证

图 8.5 所示为相移量分别为 0 、π/2 、π 和 3π/2 的 4 幅离面位移等值条纹图。

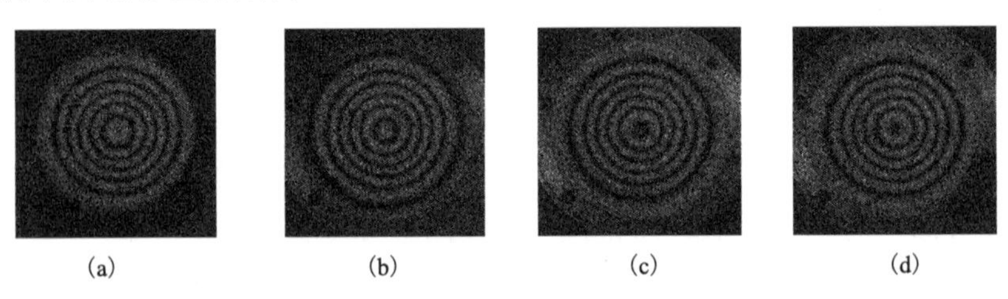

图 8.5　离面位移等值条纹

图 8.6 所示为离面位移相图，其中图 8.6(a)是 $0\sim2\pi$ 范围内的包裹相位分布；图 8.6(b)是连续相位分布。

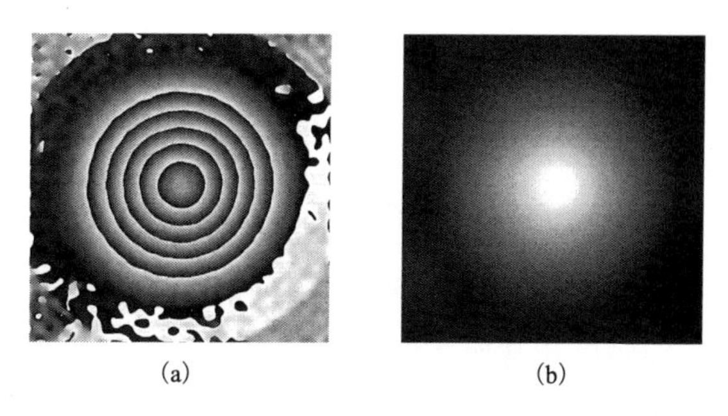

图 8.6 离面位移相位图

8.2 数字散斑剪切干涉

数字散斑剪切干涉系统如图 8.7 所示。物体由准直激光照射，物面散射光聚焦于 CCD 相机的成像靶面。通过倾斜其中的一块平面反射镜引起像面上两个散斑场相互错位，两个剪切散斑场相干叠加产生合成散斑场。

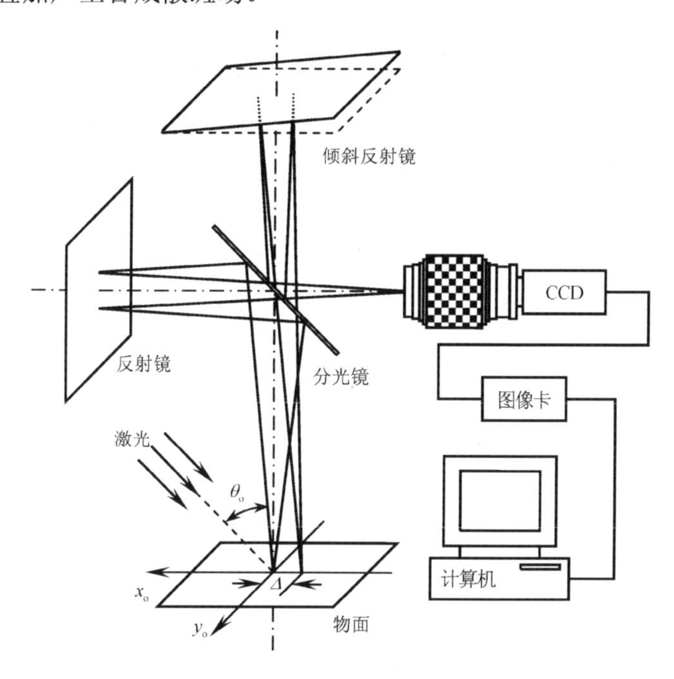

图 8.7 数字散斑剪切干涉系统

1. 条纹形成

物体变形前 CCD 记录的强度分布为

$$I_1 = I_{o1} + I_{o2} + 2\sqrt{I_{o1}I_{o2}}\cos\varphi \tag{8.26}$$

式中，I_{o1} 和 I_{o2} 分别为对应于两个剪切散斑场的强度分布；φ 为两个散斑场之间的相位差。

同理，物体变形后 CCD 记录的强度分布为

$$I_2 = I_{o1} + I_{o2} + 2\sqrt{I_{o1}I_{o2}}\cos(\varphi + \delta) \tag{8.27}$$

当激光垂直入射时：

$$\delta = 2k\frac{\partial w_o}{\partial x}\varDelta \tag{8.28}$$

式中，$\varDelta$ 为物面剪切量；$\partial w_o/\partial x$ 为离面位移沿 x 方向的导数(斜率)。

物体变形前后所记录的强度相减所得差的平方可表示为

$$E = (I_2 - I_1)^2 = 8I_{o1}I_{o2}\sin^2\left(\varphi + \frac{1}{2}\delta\right)(1 - \cos\delta) \tag{8.29}$$

显然，暗纹将产生于 $\delta = 2n\pi$ 处，即

$$\frac{\partial w_o}{\partial x} = \frac{n\pi}{k\varDelta} = \frac{n\lambda}{2\varDelta} \quad (n = 0, \pm 1, \pm 2, \cdots) \tag{8.30}$$

亮纹将产生于 $\delta = (2n+1)\pi$ 处，即

$$\frac{\partial w_o}{\partial x} = \frac{(2n+1)\pi}{2k\varDelta} = \frac{(2n+1)\lambda}{4\varDelta} \quad (n = 0, \pm 1, \pm 2, \cdots) \tag{8.31}$$

2. 相位分析

物体变形前首先记录一幅数字散斑图，物体变形后再记录相移量分别为 -3α、$-\alpha$、α 和 3α（α 为常数)的 4 幅数字散斑图。物体变形后的 4 幅数字散斑图与物体变形前的数字散斑图进行相减并平方后，可得到相移量分别为 -3α、$-\alpha$、α 和 3α 的四幅斜率等值条纹图。这些等值条纹图经过低通滤波后可分别表示为

$$\begin{aligned}
<E_1> &= 4<I_{o1}><I_{o2}>[1-\cos(\delta - 3\alpha)] \\
<E_2> &= 4<I_{o1}><I_{o2}>[1-\cos(\delta - \alpha)] \\
<E_3> &= 4<I_{o1}><I_{o2}>[1-\cos(\delta + \alpha)] \\
<E_4> &= 4<I_{o1}><I_{o2}>[1-\cos(\delta + 3\alpha)]
\end{aligned} \tag{8.32}$$

联立求解，可得物体变形相位分布为

$$\delta = \arctan\left[\tan\beta\frac{(<E_2>-<E_3>)+(<E_1>-<E_4>)}{(<E_2>+<E_3>)-(<E_1>+<E_4>)}\right] \tag{8.33}$$

式中，β 可通过下式得到

$$\beta = \arctan\sqrt{\frac{3(<E_2>-<E_3>)-(<E_1>-<E_4>)}{(<E_2>-<E_3>)+(<E_1>-<E_4>)}} \tag{8.34}$$

由式(8.33)得到的相位仍然是包裹相位，但通过相位展开算法可转化为连续相位。利用所得的连续相位，离面位移导数可表示为

$$\frac{\partial w_o}{\partial x} = \frac{\delta_u}{2k\varDelta} = \frac{\lambda\delta_u}{4\pi\varDelta} \tag{8.35}$$

3. 实验验证

图 8.8 所示为相移量分别为 -3α 、 $-\alpha$ 、 α 和 3α 的 4 幅斜率等值条纹图。

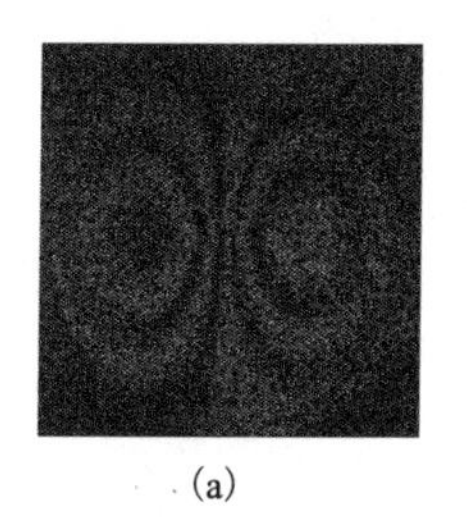
(a)

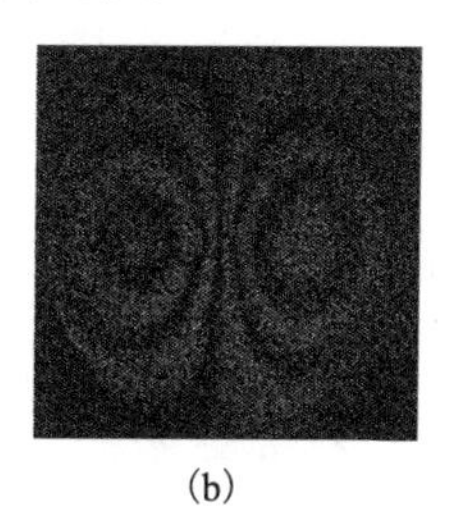
(b)

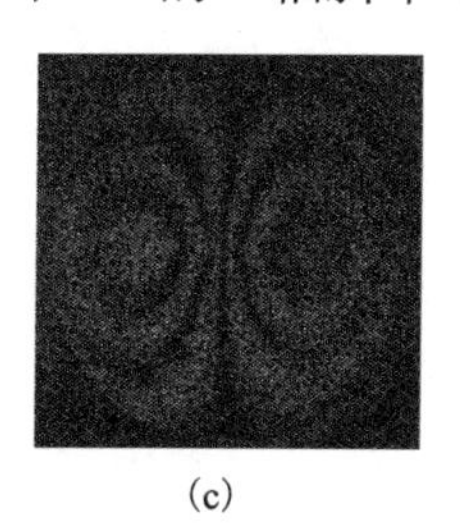
(c)

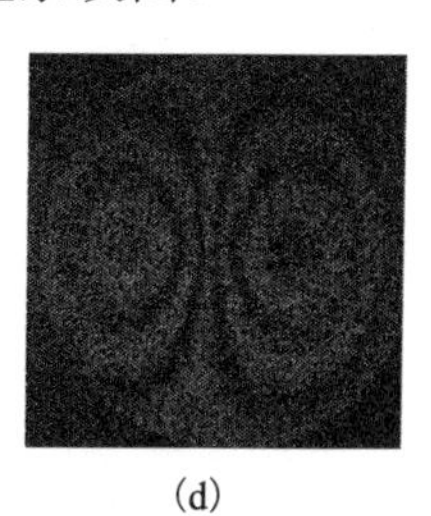
(d)

图 8.8　斜率等值条纹

图 8.9 所示为斜率相位图，其中图 8.9(a)是零相移斜率等值条纹图；图 8.9(b)是 $0\sim2\pi$ 范围内的包裹相位图；图 8.9(c)是连续相位图。

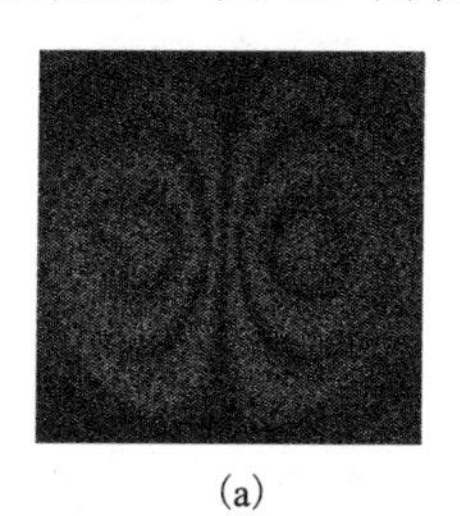
(a)

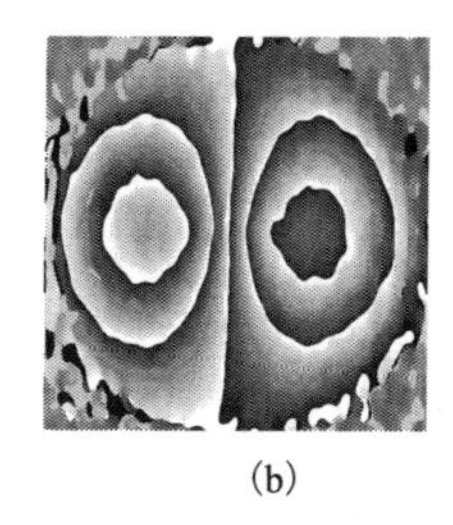
(b)

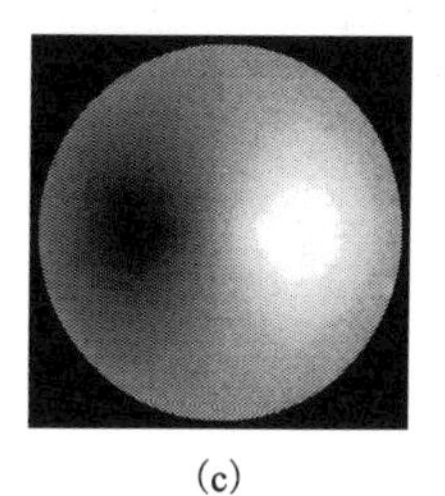
(c)

图 8.9　斜率相位图

第 9 章　数字图像相关与粒子图像测速

采样数字散斑照相记录的散斑图既可以相加存储也可以分开存储，因此数字散斑照相可以采用相加、相减或相关方法对数字散斑图进行处理以获取物体的变形信息。目前，数字散斑照相主要采用相关方法。当采用空域相关进行固体变形测量时，数字散斑照相称为数字图像相关。当采用频域相关进行流场速度测量时，数字散斑照相称为粒子图像测速。

9.1　数字图像相关

数字图像相关是指通过数码相机(如 CCD)记录物体变形前后的散斑场，利用散斑场之间的相关特性，确定相关系数的极值位置，进行物体位移或变形测量。

9.1.1　图像相关原理

散斑随机分布，即散斑场内任一点周围的散斑分布与其他点周围的散斑分布互不相同，因此散斑场内以某点为中心的子区可作为载体。根据该子区在变形前后散斑场中的移动和变化，便可获得该点的位移和变形。

当物体发生位移或变形时，散斑图上待测点由 $P(x,y)$ 移至 $P(x',y')$，以 $P(x,y)$ 为中心的子区由 $\Delta S(x,y)$ 变为 $\Delta S(x',y')$，如图 9.1 所示。因为 $\Delta S(x,y)$ 与 $\Delta S(x',y')$ 的相似程度最高，因此 $\Delta S(x,y)$ 与 $\Delta S(x',y')$ 的相关系数将达到最大值。根据相关系数峰值位置即可确定子区位移。

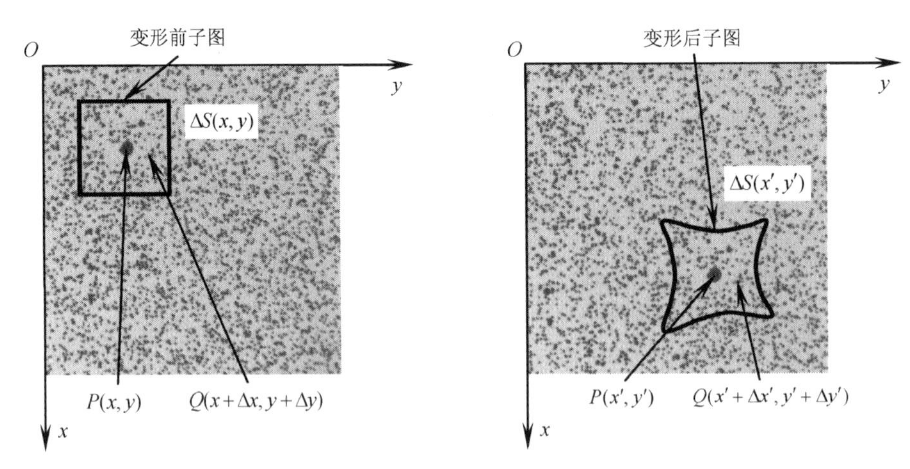

图 9.1　图像相关原理

当子区中心由 $P(x,y)$ 移至 $P(x',y')$ 时，有

$$x' = x + u(x,y),\quad y' = y + v(x,y) \tag{9.1}$$

式中，$u(x,y)$ 和 $v(x,y)$ 为子区中心分别在 x 和 y 方向的位移分量。

考虑子区内 $P(x,y)$ 的相邻点 $Q(x+\Delta x, y+\Delta y)$，其中 Δx 和 Δy 为变形前 $Q(x+\Delta x, y+\Delta y)$ 与 $P(x,y)$ 分别在 x 和 y 方向的距离。设相邻点在变形后由 $Q(x+\Delta x, y+\Delta y)$ 移到 $Q(x'+\Delta x', y'+\Delta y')$，其中 $\Delta x'$ 和 $\Delta y'$ 可分别表示为

$$\Delta x' = \Delta x + \Delta u(x,y), \quad \Delta y' = \Delta y + \Delta v(x,y) \tag{9.2}$$

式中，$\Delta u(x,y)$ 和 $\Delta v(x,y)$ 可表示为

$$\Delta u(x,y) = \frac{\partial u(x,y)}{\partial x}\Delta x + \frac{\partial u(x,y)}{\partial y}\Delta y, \quad \Delta v(x,y) = \frac{\partial v(x,y)}{\partial x}\Delta x + \frac{\partial v(x,y)}{\partial y}\Delta y \tag{9.3}$$

因此，当相邻点由 $Q(x+\Delta x, y+\Delta y)$ 移到 $Q(x'+\Delta x', y'+\Delta y')$ 时，所产生的位移分量可表示为

$$\begin{aligned} u(x+\Delta x) &= (x'+\Delta x') - (x+\Delta x) = u(x,y) + \frac{\partial u(x,y)}{\partial x}\Delta x + \frac{\partial u(x,y)}{\partial y}\Delta y \\ v(x+\Delta x) &= (y'+\Delta y') - (y+\Delta y) = v(x,y) + \frac{\partial v(x,y)}{\partial x}\Delta x + \frac{\partial v(x,y)}{\partial y}\Delta y \end{aligned} \tag{9.4}$$

显然，子区内任何点的位移都可以通过子区中心的位移分量 u 和 v 及其导数 $\partial u/\partial x$ 、$\partial u/\partial y$ 、$\partial v/\partial x$ 和 $\partial v/\partial y$ 来表示。

9.1.2　图像相关算法

1. 自相关法

对于高速运动的物体，通常需要把物体的两个瞬时散斑场存储到相同帧存，采用自相关法确定散斑场的位移分布。当散斑图内各点位移的大小和方向互不相同时，需要把散斑图分为多个子区。在同一子区内散斑具有近似相同的位移。双曝光子区的强度分布可表示为

$$I(m,n) = I_1(m,n) + I_2(m,n) \tag{9.5}$$

式中，$m = 0,1,2,\cdots,K-1$；$n = 0,1,2,\cdots,L-1$；$K \times L$ 表示子区尺寸。采用自相关法，双曝光子区的相关系数分布可表示为

$$R = (I - \langle I \rangle) \circ (I - \langle I \rangle) \tag{9.6}$$

式中，$\langle \cdot \rangle$ 表示取平均；$\circ$ 表示相关运算。由于两个函数卷积的傅里叶变换等于两个函数傅里叶变换的乘积，因此式(9.6)还可表示为

$$R = \mathrm{IFT}\{\mathrm{FT}\{I\}\mathrm{FT}\{\mathrm{RPI}\{I^*\}\}\} \tag{9.7}$$

式中，FT 和 IFT 分别表示傅里叶变换和逆变换；RPI 表示图像旋转 π；* 表示复共轭。在实际应用中，通常都是采用快速傅里叶变换进行相关计算，因为这种方法比直接计算相关要快得多。

图 9.2 所示为自相关法的分析结果，其中图 9.2(a) 是双曝光散斑图；图 9.2(b) 是相关系数分布。

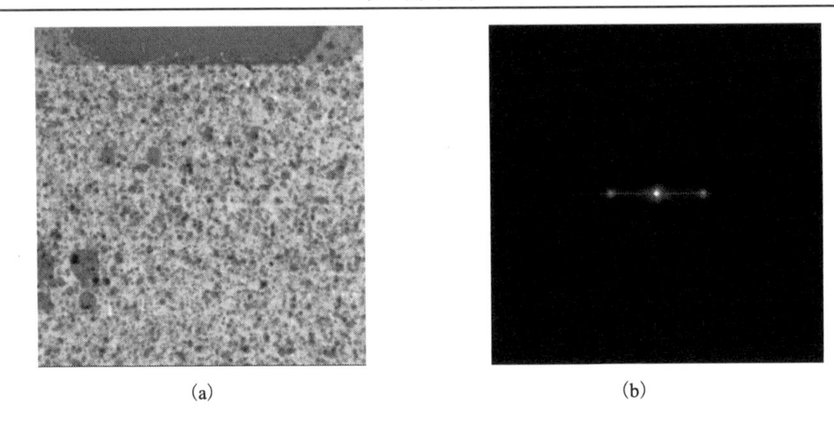

(a)　　(b)

图 9.2　自相关结果

由图 9.2 可见，采用自相关法，相关系数分布会出现 3 个峰。中心峰与两侧峰之间的距离即为子区中心位移的大小，但子区中心位移的方向无法确定。

2. 互相关法

如果可能，通常都把物体变形前后的散斑图分开存储，可采用互相关法确定散斑位移。同样，当散斑图内各点位移的大小和方向互不相同时，也需要把散斑图分为多个子区，在同一子区内，散斑具有近似相同的位移。

设子区在变形前后的强度分布分别由 $I_1(m,n)$ 和 $I_2(m,n)$ 表示，其中，$m=0,1,2,\cdots,K-1$；$n=0,1,2,\cdots,L-1$。采用互相关法，变形前后子区的相关系数可表示为

$$R=(I_2-<I_2>)\circ(I_1-<I_1>) \tag{9.8}$$

同理，式(9.8)也可表示为

$$R=\mathrm{IFT}\{\mathrm{FT}\{I_2\}\mathrm{FT}\{\mathrm{RPI}\{I_1^*\}\}\} \tag{9.9}$$

图 9.3 所示为互相关法的分析结果，其中图 9.3(a)和图 9.3(b)分别是变形前后的散斑图；图 9.3(c)是相关系数分布。由图 9.3 可见，采用互相关法，相关系数分布只出现 1 个峰，该峰相对子区中心的距离和方向即为子区中心位移的大小和方向。

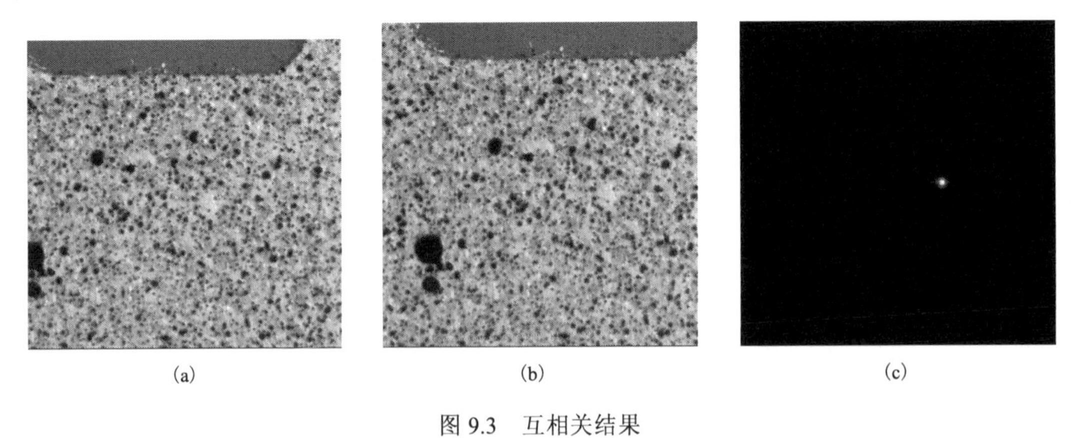

(a)　　(b)　　(c)

图 9.3　互相关结果

9.1.3　图像相关系统

图像相关系统如图 9.4 所示。用白光(或激光)照射试件。为使试件表面光场均匀，可以采用对称光源。

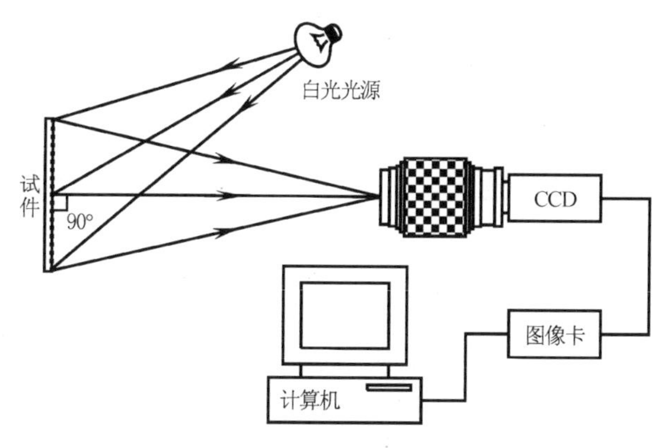

图 9.4　图像相关系统

9.1.4　图像相关实验

图 9.5 所示为物体变形前后的白光数字散斑图。

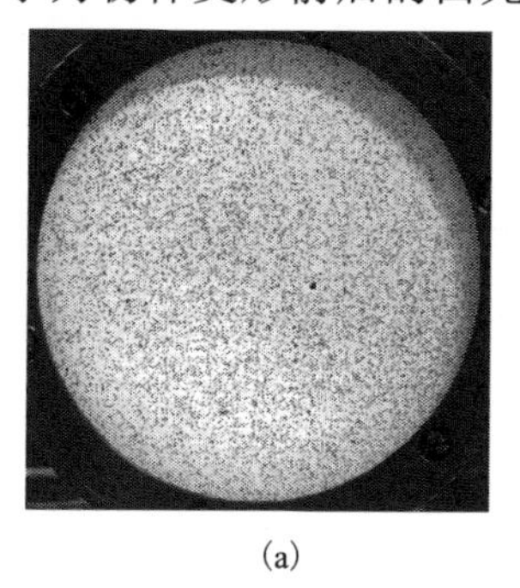

(a)

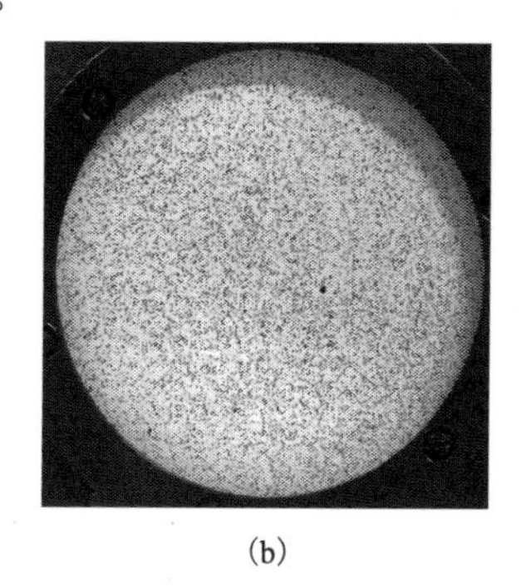

(b)

图 9.5　白光数字散斑图

图 9.6 所示为位移分量。图 9.6(a)表示竖直分量(向下为正)；图 9.6(b)表示水平分量(向右为正)。

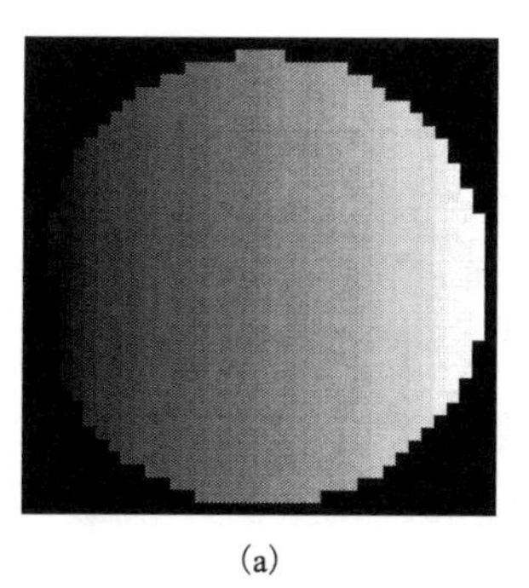

(a)

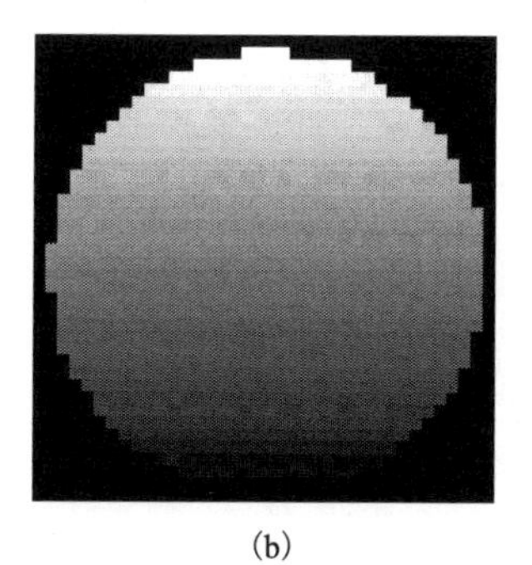

(b)

图 9.6　位移分量

图 9.7 所示为位移大小和方向。图 9.7(a)表示大小；图 9.7(b)表示方向。

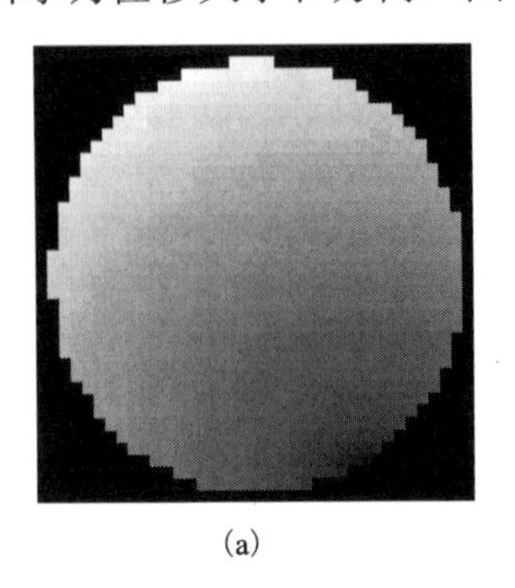
(a)

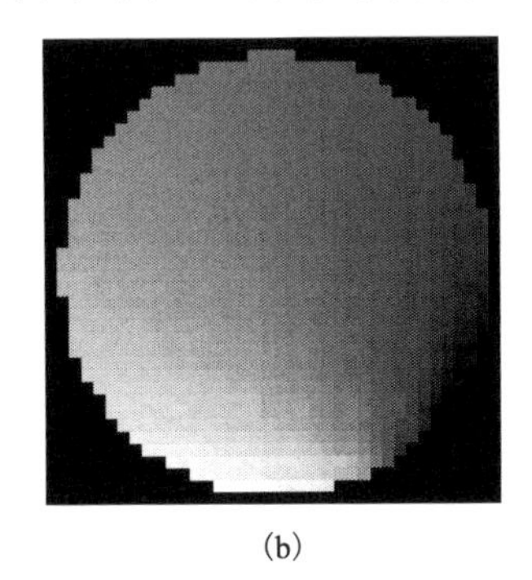
(b)

图 9.7　位移大小和方向

9.2　粒子图像测速

粒子图像测速是基于散斑照相而提出的高精度非接触全场流速测量技术。通过测量流场中示踪粒子的运动而获得流场速度分布。

粒子图像测速技术具有高精度、非接触和全场测量等优点，因而目前已应用于各种流场测量。

9.2.1　图像测速原理

如图 9.8 所示，设在时刻t和$t+\Delta t$，流场中某一示踪粒子P分别处于位置$[x(t), y(t)]$和$[x(t+\Delta t), y(t+\Delta t)]$，则在时间间隔$\Delta t$内示踪粒子$P$在$x$和$y$方向的位移分量可表示为

$$u(t) = x(t+\Delta t) - x(t), \quad v(t) = y(t+\Delta t) - y(t) \tag{9.10}$$

因此，在时间间隔Δt内，示踪粒子P在x和y方向的速度分量可表示为

$$v_x = \frac{u(t)}{\Delta t}, \quad v_y = \frac{v(t)}{\Delta t} \tag{9.11}$$

通常，时间间隔Δt很短，因此式(9.11)求得的速度可以看成时刻t的瞬时速度分量。

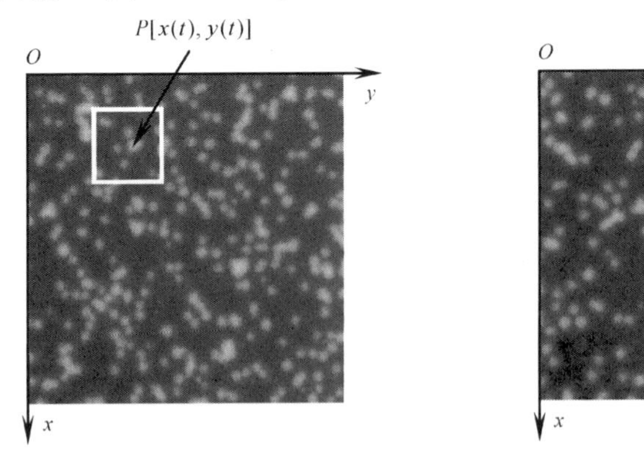

图 9.8　图像测速原理

9.2.2　图像测速算法

1. 自相关法

把两个不同时刻的粒子图像记录在同一帧存中，则双曝光子区的强度分布可以表示为

$$I(m,n)=I_1(m,n)+I_2(m,n) \tag{9.12}$$

式中，$m=0,1,2,\cdots,K-1$；$n=0,1,2,\cdots,L-1$；$K\times L$ 为子区尺寸；$I_1(m,n)$ 和 $I_2(m,n)$ 表示两个不同时刻的强度分布。当子区足够小时，可认为子区中的粒子具有相同的位移(或速度)，则有

$$I_2(m,n)=I_1(m-u,n-v) \tag{9.13}$$

式中，$u=u(m,n)$ 和 $v=v(m,n)$ 为子区粒子分别在 x 和 y 方向的位移分量。对式(9.12)进行傅里叶变换，则有

$$A(f_x,f_y)=\mathrm{FT}\{I(m,n)\}=\mathrm{FT}\{I_1(m,n)\}+\mathrm{FT}\{I_2(m,n)\} \tag{9.14}$$

式中，f_x 和 f_y 为变换面上分别沿 x 和 y 方向的离散频率坐标。利用傅里叶变换的平移性质，有

$$\mathrm{FT}\{I_2(m,n)\}=\mathrm{FT}\{I_1(m-u,n-v)\}=\mathrm{FT}\{I_1(m,n)\}\exp[-\mathrm{i}2\pi(uf_x+vf_y)] \tag{9.15}$$

利用式(9.15)，则式(9.14)可表示为

$$A(f_x,f_y)=\mathrm{FT}\{I_1(m,n)\}\{1+\exp[-\mathrm{i}2\pi(uf_x+vf_y)]\} \tag{9.16}$$

因此，变换面上的强度分布可表示为

$$I(f_x,f_y)=|A(f_x,f_y)|^2=2|\mathrm{FT}\{I_1(m,n)\}|^2\{1+\cos[2\pi(uf_x+vf_y)]\} \tag{9.17}$$

显然，在变换面可观察到余弦干涉条纹。

对式(9.16)进行自相关运算，或对式(9.17)进行傅里叶变换，可得子区的相关系数分布为

$$\begin{aligned}R(m,n)=\mathrm{FT}\{|A(f_x,f_y)|^2\}&=2\mathrm{FT}\{|\mathrm{FT}\{I_1(m,n)\}|^2\}+\mathrm{FT}\{|\mathrm{FT}\{I_1(m+v,m+v)\}|^2\}\\&\quad+\mathrm{FT}\{|\mathrm{FT}\{I_1(m-u,m-v)\}|^2\}\end{aligned} \tag{9.18}$$

显然，子区的相关系数分布将出现 3 个峰，因此测量子区位移就归结为寻找中心峰与两侧峰之间的距离，但位移指向通过自相关法无法确定。

图 9.9 所示为自相关法分析结果，其中图 9.9(a)是双曝光散斑图，图 9.9(b)是相关系数分布。

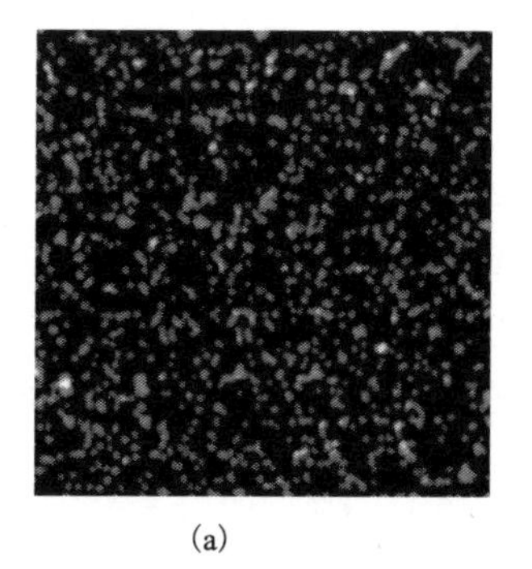

(a)

(b)

图 9.9　自相关结果

2. 互相关法

设两个不同时刻记录的粒子图像的强度分布分别由 $I_1(m,n)$ 和 $I_2(m,n)$ 表示，并设子区内粒子具有相同的位移(或速度)，则有

$$I_2(m,n)=I_1(m-u,n-v) \tag{9.19}$$

对两幅粒子子区分别进行傅里叶变换，则有

$$A_1(f_x,f_y)=\mathrm{FT}\{I_1(m,n)\},\quad A_2(f_x,f_y)=\mathrm{FT}\{I_1(m,n)\}\exp[-\mathrm{i}2\pi(uf_x+vf_y)] \tag{9.20}$$

对式(9.20)进行互相关运算，得子区的相关系数分布为

$$R(m,n)=\mathrm{FT}\{[A_1(f_x,f_y)][A_2(f_x,f_y)]^*\}=\mathrm{FT}\{|\mathrm{FT}\{I_1(m-u,n-v)\}|^2\} \tag{9.21}$$

显然，采用互相关法，子区的相关系数分布仅出现 1 个峰，该峰相对子区中心的距离和方向即为子区的位移大小和方向。

图 9.10 所示为互相关法的分析结果，其中，图 9.10(a)和图 9.10(b)分别是变形前后的散斑图；图 9.10(c)是相关系数分布。

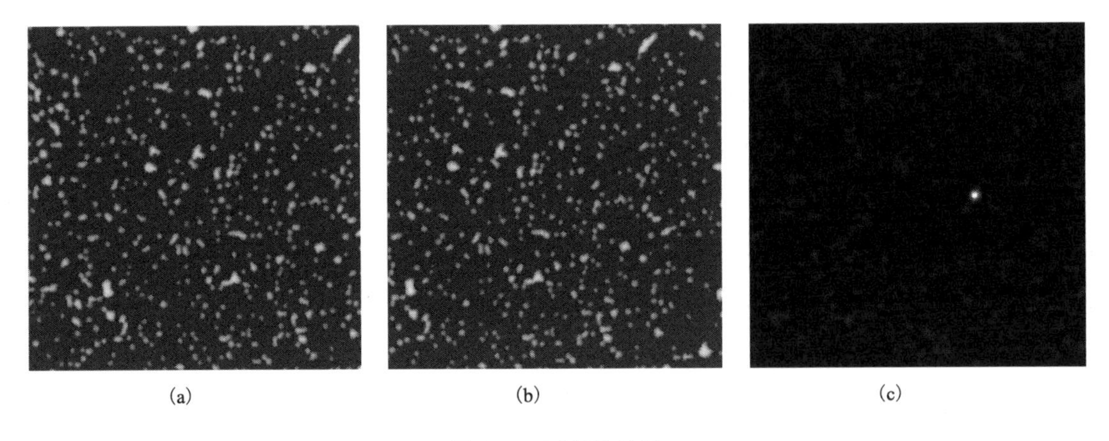

(a)　(b)　(c)

图 9.10　互相关结果

9.2.3　图像测速系统

图像测速系统如图 9.11 所示。激光器发出的脉冲激光通过柱面镜形成激光片光，照射所要研究的流场。当激光片光照射到待测流场中的示踪粒子时，激光片光将在粒子上发生散射，在激光片光所在平面的法线方向由 CCD 记录粒子图像。经过两次或多次曝光，不同时刻的粒子图像被分别存储到计算机中。通过自相关或互相关运算，即可根据已知时间间隔内的子区位移，计算出流场的速度分布。

粒子图像测速系统主要包括示踪粒子、激光片光、同步装置、图像采集与处理系统等。

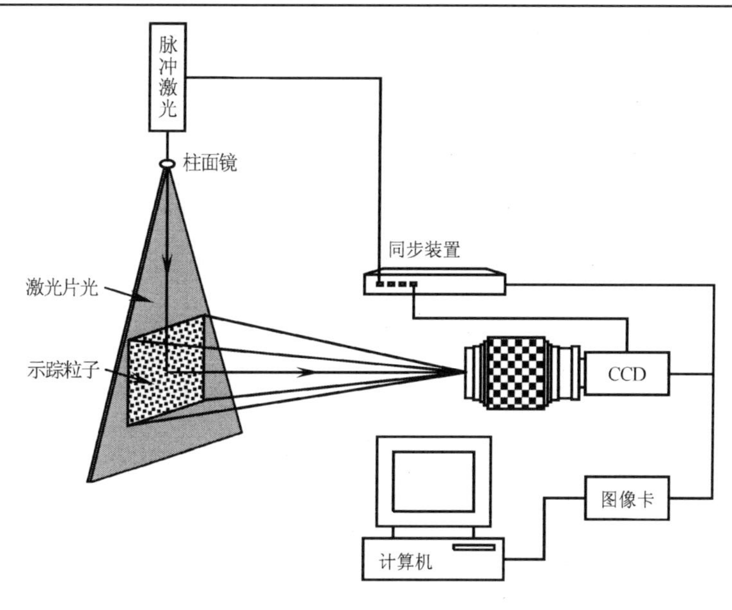

图 9.11　图像测速系统

1. 示踪粒子

采用粒子图像测速技术进行流速测量时，需要在流场中投入跟随性和散射性良好且密度与流体相当的示踪粒子。示踪粒子直接反映流场特性，它的选取及使用是粒子图像测速技术的关键。示踪粒子必须有良好的跟随性和散射性，浓度也要适当。

示踪粒子的跟随性是影响测量精度的重要因素，作为反映流场速度的中间物，示踪粒子必须能够很好地跟随流体，也就是说粒子的跟随性要好。示踪粒子跟随流体的程度不仅取决于示踪粒子粒径、粒子浓度等，还取决于流体带动粒子的能力，如流体的运动状态、流体黏度、流体密度等。总体来说，如果粒子太大则自身的惯性很大，流场将不能完全带动粒子，使得粒子的运动与流场的运动相差较大；如果粒子太小，则很容易受到外界的干扰而不能反映流场的真实状态。

粒子的散射性能将影响图像的分辨率。为了提高分辨率，通常可以增加激光光源功率和提高粒子的散射性能，但增加激光光源功率的代价太大，在激光功率一定的情况下，人们更倾向于提高粒子的散射性能。粒子的大小、形状、粒子折射率和流体折射率等都将影响粒子的散射性能，另外，接收照射光的方向存在最佳角度，测速时可以进行适当的调节以提高分辨率。

在流速测量中需要考虑粒子的浓度问题。当浓度太高时，粒子就会重叠在一起；当粒子浓度太低时，就会得不到足够多的速度矢量，从而不能反映流场的全场速度分布。目前常用的示踪粒子有很多种，适用于水的示踪粒子有荧光粒子、表面镀银的空心玻璃球粒子、乳化空气泡粒子和液晶粒子等；适用于气体的有粉末粒子、二氧化钛粒子以及雾化油滴等。

2. 激光片光

粒子图像测速对所用的激光光源有一定的要求：首先是功率要高，片光要能照亮流场，使所研究的流场区域内粒子的散射光有足够的散射强度，以便记录到清晰的粒子图像；其次要能形成脉冲激光片光，利用脉冲片光将两个瞬时的流场记录下来。

激光光源通常分为两类：一类是连续激光，连续激光光源前需要附加机械式或光电式频闪装置，氩离子激光器发出的是连续光，通常用于低速液态流场测量；另一类是脉冲激光，这种激光器的每个振荡器和放大器都可分别触发，因此可无限小地控制激光脉冲间隔和单独调节单个激光脉冲的能量。一般在粒子图像测速系统中采用两台脉冲激光器，用外同步控制装置来分别触发激光器以产生脉冲。

粒子图像测速需要采用激光片光照射，激光脉冲的片光由柱面镜和球面镜联合产生，准直了的光束通过柱面镜后形成激光片光，同时球面镜用于控制片光的厚度。

3. 同步装置

同步控制器用来协调粒子图像测速系统各个部分的工作时序，由计算机进行控制。它控制脉冲发出和图像采集的顺序，通过内部产生周期性的脉冲触发信号，经过多个延时通道同时产生多个经过延时的触发信号，用来控制激光器、CCD和图像采集卡，使它们工作在严格同步的信号基础上，保证各部分按一定的时间顺序协调工作。计算机用于存储图像卡提供的图像数据，通过粒子图像测速系统软件可以实时完成速度场的计算、显示和存储。

同步控制器提供周期的外触发信号，激光器的两个氙灯经过一定的时间延时，间隔点亮发光，当氙灯发光强度达到最大峰值时，经过适当延时的两路Q开关被同步控制器提供的延时信号触发，激光器发出具有一定时间间隔的双脉冲激光，同时，CCD也收到同步控制器提供的触发信号。使用软件设定第一幅图像的曝光时间，使激光器发出的脉冲落在第一次曝光时间内；然后经过软件设定的跨帧时间，CCD可以进行第二次曝光，这时捕捉到的是激光器的第二个脉冲。这样就实现了 CCD 触发一次得到两帧图像、捕捉双脉冲激光的功能。缩小激光器双脉冲之间的时间间隔，就可以拍摄高速运动的流场图像，计算相应的速度场。

4. 图像采集与处理系统

粒子图像采集系统是粒子图像测速系统的关键部分，它包括高分辨率 CCD 和数据采集卡。CCD的图像采集方式分成两类：一类是自相关模式，两个瞬时的粒子图像存储在相同帧存中；另一类是互相关模式，两个瞬时的粒子图像存储到不同帧存中。

粒子图像是一系列随时间变化的数字图像，这些序列图像的处理方法主要采用相关方法。相关方法分为自相关法和互相关法。

对于自相关分析，自相关系数分布有3个峰：一个中间自相关峰和它旁边的两个互相关峰。互相关峰的位置对应粒子图像的位移。自相关法的对称性造成速度方向不能自动判别。另外，自相关法的速度测量范围不大，因为进行自相关运算时，子区只能在自身图像范围内寻找最大相关系数。

对于互相关分析，互相关系数分布只有1个峰，因而可以自动判别速度方向。互相关处理时，图像中的子区在另一幅图像中寻找最大相关系数，因此互相关法的速度测量范围要比自相关法大得多。

参 考 文 献

戴福隆, 方萃长, 刘先龙, 等. 1990. 现代光测力学. 北京: 科学出版社.

戴福隆, 沈观林, 谢惠民, 等. 2010. 实验力学. 北京: 清华大学出版社.

董守荣. 1987. 波动光学. 武汉: 华中理工大学出版社.

韩军, 刘钧. 2007. 工程光学. 西安: 西安电子科技大学出版社.

黄婉云. 1985. 傅里叶光学教程. 北京: 北京师范大学出版社.

梁柱. 2005. 光学原理教程. 北京: 北京航空航天大学出版社.

刘培森. 1987. 散斑统计光学基础. 北京: 科学出版社.

王开福. 2013. 现代光测及其图像处理. 北京: 科学出版社.

王开福, 高明慧. 2010. 散斑计量. 北京: 北京理工大学出版社.

王开福, 高明慧, 周克印. 2009. 现代光测力学技术. 哈尔滨: 哈尔滨工业大学出版社.

吴健, 严高师. 2007. 光学原理教程. 北京: 国防工业出版社.

谢建平, 明海, 王沛. 2006. 近代光学基础. 北京: 高等教育出版社.

于起峰, 伏思华. 2007. 基于条纹方向和条纹等值线的ESPI与InSAR干涉条纹图处理方法. 北京: 科学出版社.

于起峰, 尚洋. 2009. 摄像测量学原理与应用研究. 北京: 科学出版社.

张如一, 陆耀桢. 1981. 实验应力分析. 北京: 机械工业出版社.

赵清澄, 石沅. 1987. 实验应力分析. 北京: 科学出版社.